ANIMAL DRUGS and HUMAN HEALTH

HOW TO ORDER THIS BOOK

BY PHONE: 800-233-9936 or 717-291-5609, 8AM–5PM Eastern Time

BY FAX: 717-295-4538

BY MAIL: Order Department
Technomic Publishing Company, Inc.
851 New Holland Avenue, Box 3535
Lancaster, PA 17604, U.S.A.

BY CREDIT CARD: American Express, VISA, MasterCard

ANIMAL DRUGS and HUMAN HEALTH

EDITED BY

Lester M. Crawford, D.V.M., Ph.D.
EXECUTIVE DIRECTOR
ASSOCIATION OF AMERICAN VETERINARY MEDICAL COLLEGES
WASHINGTON, D.C.

Don A. Franco, D.V.M., M.P.H., Dipl. A.C.V.P.M.
DIRECTOR, SCIENTIFIC SERVICES, NATIONAL RENDERERS ASSOCIATION
and
DIRECTOR, SCIENTIFIC AFFAIRS, ANIMAL PROTEIN PRODUCERS INDUSTRY
ALEXANDRIA, VA

TECHNOMIC
PUBLISHING CO., INC.
LANCASTER · BASEL

Animal Drugs and Human Health

a **TECHNOMIC**®publication

Published in the Western Hemisphere by
Technomic Publishing Company, Inc.
851 New Holland Avenue, Box 3535
Lancaster, Pennsylvania 17604 U.S.A.

Distributed in the Rest of the World by
Technomic Publishing AG
Missionsstrasse 44
CH-4055 Basel, Switzerland

Printed in the United States of America

10 9 8 7 6 5 4 3 2 1

Main entry under title:
  Animal Drugs and Human Health

A Technomic Publishing Company book
Bibliography: p.

Library of Congress Catalog Card No. 93-61759
ISBN No. 1-56676-102-6

Table of Contents

Foreword

THESE chapters reviewing the issue of drug and chemical residues in meat and other animal products is timely indeed. At no time in history have scientists and society been more concerned with questions about contamination of the food supply by chemical residues, either from intentional and focused uses, or from inadvertent introduction.

On the other hand, the technological advances in production of animal-derived foods achieved during the last 30 years have been phenomenal. Much of this progress has been due to the development of innovative pharmaceutical preparations and chemical pesticides. However, these advances have not been without public health concerns. The chemical products involved, and the public concerns expressed among developed countries, are shared in many parts of the world, and the problems may be magnified in developing regions.

This global aspect of chemical residues argues for uniform international standards to protect the public health. A well-informed public recognizes that decisions made by governments and industry leaders, who manage the risks involved in these products, will affect not only the abundance of food, but the safety of this food. Therefore, these societal decisions must be based on the finest, most exquisite science possible. Further, since yesterday's decisions may not be valid for today, responsible decision makers must make certain that knowledge — and our ability to integrate it into effective risk assessment and risk management decisions — advances over time.

This demand on world leadership, especially on regulatory officials, requires an awareness of the worldwide scientific literature in this area, as well as a close association with counterpart officials throughout the world. International cooperation in the risk assessment and risk management of animal drugs and pesticides, so that standard-setting can be accomplished through worldwide, scientific consultation, is essential. A safe food supply is the primary goal of us all. This book is a major contribution toward accomplishing this goal.

GERALD B. GUEST, D.V.M.
Director
Center for Veterinary Medicine
U.S. Food and Drug Administration

Acknowledgements

T HE editors are very appreciative of the comments of colleagues. Special recognition is due Dr. Ronald E. Engel for his painstaking effort and extensive review of manuscripts. His support contributed substantially to the completion of the book. Edith Kennard was an outstanding resource, and willingly assisted in every conceivable way. To these two people, we are most grateful.

L. M. CRAWFORD
D. A. FRANCO

CHAPTER 1

The Public Health Perspective

LESTER M. CRAWFORD, D.V.M., Ph.D.[1]
DON A. FRANCO, D.V.M., M.P.H., Dipl. A.C.V.P.M.[2]

1.1 INTRODUCTION

IT has been universally accepted that the discovery of antimicrobial agents was a major and significant milestone in the history of medicine, and in human and animal health (Nichols and Keys, 1984). These powerful drugs were initially considered miraculous by the medical and veterinary professions, because their use saved lives from severe infections formerly unresponsive to other forms of traditional therapy. Thus, the advent of antimicrobial agents was readily accepted as the ultimate. This was followed by the conviction of researchers and the pharmaceutical industry that the potential for developing improved and more potent antimicrobials had no barriers. At the same time, the use of these miraculous drugs went beyond the standard therapeutic regimen. This diversion was particularly widespread in livestock agriculture. Antibiotics enjoyed wide subtherapeutic use in animal feeds for growth promotion, improvement of feed efficiency, and disease prophylaxis (NRC, 1980).

Subtherapeutic use in animal husbandry has increased steadily since 1950, and in the latter part of the 1970s approximately half of all antibiotics produced were destined for addition to animal feeds, or for other agricultural use (NRC, 1980). Research data started to demonstrate findings that the use of antimicrobials in animal feed increased the prevalence of R+ enteric organisms. The implication was that some of these organisms may be pathogenic to humans. This was further authenticated by research con-

[1]Executive Director, Association of American Veterinary Medical Colleges, Washington, D.C., USA.
[2]Director, Scientific Services, National Renderers Association, and Director, Scientific Affairs, Animal Protein Producers Industry, Alexandria, Virginia, USA.

firmation that persons in close contact with animals receiving antimicrobials are more likely to harbor antimicrobial-resistant *E. coli* than nonexposed persons (Bland, 1987).

Traditionally, the major regulatory concerns are to determine the adverse effects on human health. Producers and the industry have demonstrated heightened anxieties on the possible negative esthetic effects on the consuming public by the presence of chemicals in the food supply (Beran, 1987). There is an increasing public outcry to know what residues and contaminants are in our food supply, and a demand that food be free of residues or metabolites of drugs that could impact the public health.

Three out of five consumers in a survey considered antibiotics in poultry and livestock as a serious health hazard. An additional third consider antibiotic use somewhat of a hazard (Scroggins, 1987). The reality is that the use of antimicrobial agents is widespread. They have been fed at some time during life to nearly 80% of poultry, 75% of swine, 60% of feedlot cattle, and 75% of dairy calves. Estimates are that antibiotic feed supplementation for the production of beef, pork and poultry has saved producers and, ultimately, the consuming public in excess of $3.5 billion annually (ASTC, 1981; Marshall, 1980). The dilemma of use, misuse, and the potential public health hazards will not abate until the uncertainty of the existing data and future research provides the finite scientific experimentation and proof. Currently, researchers recommend "continued monitoring and occasional review of the possible effects on humans resulting from the subtherapeutic use of antimicrobials in animal feeds" (NRC, 1980).

Pesticides are chemicals used throughout the agricultural world to destroy or control weeds, fungi, insects, and other pests. Some tend to remain in food as residues; some are not properly applied according to established guidelines/instructions and the resulting residues due to misuse can pose potential health risks to consumers (U.S. Congress, 1988).

Public concern over the use of pesticides and the resultant potential residues has increased dramatically during the last decade. A 1988 national survey by the Food Marketing Institute readily demonstrated the existing apprehension. In this poll, 75% of the surveyed were very concerned about pesticides in food. This was both a surprising statistic and interesting perspective, because it was higher than those who worried about cholesterol, fats, salts, additives, or other components used in food (FMI, 1988). Assumptions were somewhat corrected and modified by these survey findings, because the prevalent conviction at the time was that the public at large was preoccupied by fat consumption and cholesterol.

The use of hormones in food-producing animals is generally limited to anabolic implants. The latter are growth regulators that repartition or *shift* growth. Their use is confined to food-producing ruminants (cattle, calves, sheep), in which they cause the shift from fat to lean, an increase in growth rate, and reduced feed requirements for growth. The implants are used with efficacious results in every phase of beef production and in sheep feedlots (USDA, 1987).

Anabolic agents are usually classified in four major categories:

- steroid hormones
- synthetic hormones
- xenobiotic compounds
- other growth-promoting compounds

Questions about the use of hormones (naturally-occurring sex steroids like estradiol and testosterone, or syntheric hormones like trenbolone acetate and zeranol) arise from concerns about whether or not they produce any harmful effects on consumers who eat meat from treated animals. The United States Food and Drug Administration (FDA) has maintained that hormones approved for use (estradiol, progesterone, testosterone, and zeranol) are safe when used according to label directions, and that no physiologic effect could be expected in consumers eating meat.

1.2 THE PUBLIC HEALTH IMPLICATIONS ASSOCIATED WITH ANTIBIOTICS, PESTICIDES AND HORMONES

The public health implications will be limited to synoptic concepts and a highlight of the major aspects of antibiotic and hormonal use, plus the current concerns with pesticide use and contamination.

1.2.1 ANTIBIOTICS

The current mechanisms to determine whether human health is affected by the subtherapeutic or therapeutic antibiotic use in animals is stymied by the diverse opinions associated with defining exactly what constitutes a hazard to human health. Nonetheless, this aspect of antiobiotic use in food production will be approached from the public-health perspective regardless of obvious limitations.

Several well-documented studies have implicated residues in foods as agents of immediate, delayed, or chronic allergic reactions. These reactions are preceded by a sensitization period (Kilara, 1982). A second, but most significant, consideration relative to antibiotic use pertains to the complex of microorganisms which—regardless of the host, route of administration, or levels of therapy—is the resulting appearance of resistant populations of bacteria that are causally related both to disease and to resident flora (Murray, 1984). This antimicrobial drug resistance is commonly mediated through R + plasmids which may acquire or transfer resistance genes on transposable elements to chromosomes (Beran, 1987). There is a potential for these resistant bacteria in food-producing animals to be transferred to humans. Epidemiological evidence has proven that R+ plasmids are very common in the intestinal flora of different classes of livestock, and that people working in the same environment, and in close contact with these animals, do pick up these drug-resistant bacteria (Beran, 1987).

The first two problems have direct pertinence to public health. The third concern can be classified under a twofold consideration, a mixture of technoeconomic and zoonotic, but with secondary public health importance. The diseases of dairy cattle, especially the various manifestations of acute and chronic mastitis, are predominantly treated with antibiotics (single or combined) and other medications. This often results in measurable amounts of antibiotics in the milk. The use of this contaminated milk to produce dairy products will result in residues in these products that could survive both pasteurization and freezing (Kilara, 1982). Characteristic of many outbreaks of salmonellosis was the use of antibiotics in affected animals, which eventually served as sources of transmission to people. Persons on concomitant antibiotic therapy at the time of exposure to the animal or its products tended to be more susceptible to infection than persons not receiving antibiotics (Beran, 1987).

The importance and relative medical and epidemiologic significance of the possible effects on public health due to the therapeutic use of antibiotics has been the subject of numerous scientific debates and excellent studies by distinguished world scientists. Doubtless, the quest for the answers in our efforts to safeguard the public health will force us to work beyond the current limitations imposed by incomplete knowledge and follow the scientific path of absolute certainty.

We would be remiss if we discussed the antimicrobials and totally accentuated the antibiotics without a brief mention of sulfonamides, the synthetic antimicrobial agents used predominantly in swine, especially as sulfamethazine. The Food Safety and Inspection Service (FSIS) use of the sulfa-on-site (SOS) test in slaughter establishments has greatly reduced the prevalence of sulfa residues in swine. Pork that is contaminated with sulfonamide residues may cause hypersensitive reactions in sensitized humans.

1.2.2 PESTICIDES

Pesticides properly used are extraordinarily valuable to agriculture. They help control insects, weeds, and many plant fungi. Improperly used, however, pesticides can be serious hazards to public health, and they are a major public concern because of their potential for the environmental contamination of food.

The environmental contamination of food can occur in any modern society with the technical capability to produce and consume huge amounts of various chemical substances, some of which are toxic. Current estimates show approximately 70,000 chemicals in commercial production in the United States. Additional data indicate that about 50 or more of these chemicals are produced in amounts greater than 1.3 billion pounds per year (U.S. Congress, 1988). Published records show that 7% of the country's gross national product (GNP) is generated by the manufacture and distribution of chemicals totaling $113 billion annually (Storclz, 1978). These figures were based on available information for the late 1970s. Assumptions are that current production amounts and the gross national product percentage have increased greatly in the 80s and will continue into the 90s. This is a very impressive accomplishment for the chemical industry, and an interesting statistic for decision makers of regulatory agencies and the political network charged to protect consumers and the *open environment*—the chemical and biological cycles that move, transform, and remove substances from air, water, and soil (Chanlett, 1979). During the production phase, use, and disposal of these substances (many of which are pesticides), there are opportunities for losses into the environment and a resultant contamination of the food supply.

Pesticides are traditionally considered the only agricultural chemicals known to be environmental contaminants in food. Livestock, poultry and fish are usually contaminated when the production or application of pesticides occurs in the vicinity of, or the transport of residues through, the environment. Warehouses used for storing food, railroad cars, ships, or trucks that are improperly fumigated, and other food storage facilities that are not properly aired after fumigation or spraying, are obvious sources of contamination of food or feed (U.S. Congress, 1988). Manufacturing processes resulting in gases, liquid effluents, and sludges that enter the environment by air, soil and fertilization of land are also potential sources of food contamination.

Many of the chlorinated hydrocarbon pesticides like DDT, dieldrin, heptachlor and aldrin that were in vogue for many years have now been replaced by compounds that are less persistent. Some of the pesticides of the past were so chemically stable they did not decompose for years, and often biomagnified in tissue. The chemical and biological properties of the new pesticides make them unlikely to persist or accumulate in the environment (Coffin et al., 1982). This will reduce the public health risks. However, caution is still the key in pesticide use. These compounds must be handled wisely to avoid accidental contamination. Label precautions must be stringently observed and followed to preclude adverse effects to food, water, and other major components of the *open environment*.

1.2.3 HORMONES

In any discussion of the use of anabolic hormones in animals, it is imperative and pertinent to recapitulate the major concerns surrounding hormone use in humans to gain insight into the past and present concerns of these agents. This review and the associated discussion will be limited to prevent overlap with subject matter discussed in greater detail in Chapter 7.

As early as 1932, research workers and epidemiologists demonstrated a causal association in human use between the endogenous sex steroids and carcinogenesis. The prevailing scientific concensus, however, based on longitudinal studies, is that, contrary to the initial fears, these agents are not carcinogenic per se, but play the role of promoters after the process has been initiated by other agents (e.g. viral, chemical, physical) (Guest et al., 1987).

Then, there is the apparently similar, but entirely different, situation with the long-term human use of diethylstilbestrol (DES) and the subsequent determination that DES was genotoxic and capable of initiating the carcinogenic process even at minute concentrations. These early findings have caused lingering doubts about hormonal use in general. This, despite the ban on the use of DES in food-producing animals over a decade ago.

The use of hormones in ruminants (cattle and sheep) has to be addressed from the public health perspective. The safety factor continues to be in the forefront in most of the important debates and policy decisions involving hormone use. International differences have split the world's scientific community on use acceptance. And at times, the impression is that emotion seems to be ruling reason, even when everyone appears to be saying the same things and making the same scientific pronouncements. In the U.S. the official position has been that the use of hormones (anabolic agents) in animals, according to label directions, is safe, and that consumers are not at risk by eating meat from treated animals. The FDA has also concluded that any enforcement of a ban on estradiol, testosterone, and progesterone will be impossible, since analytical methods cannot distinguish between those which occur naturally in food-producing animals and those that might be present from the use of these hormones as growth promoters (Miller, 1987).

Despite the European Community (EC) ban on growth-promoting hormones since 1985, many farmers throughout Europe have continued to use clenbuterol, a beta-agonist, to increase carcass weight in cattle. This is in spite of the drug's license in the EC only to treat respiratory disease in horses, specifically as a bronchodilator. The drug, which is used illegally in animal feeding, causes regression in body lipids, muscle growth, weight gain and accumulation in the liver of cattle.

There have been several reports of food poisoning related to consumption of beef livers originating from cattle fed clenbuterol. The cases typically arose in families who had consumed liver, resulting in an attack rate of 97%; no cases were seen in those who had not eaten liver. Clinically, the observed signs were muscle tremors and palpitation/tachycardia, often accompanied by nervousness, cephalgia and myalgia, lasting for about 40 hours.

1.3 SUMMARY

The environment has started to play a significant role in influencing public health planning and policy decisions. All of the established epidemiologic indicators are that the morbidity and mortality from infectious diseases continue to decline. Environmental factors, especially by ambient exposures through air, water and food, have become matters of serious concern, and now tend to occupy an increased role in public affairs and decision processes (Frank, 1985). It is apparent that decisions will often have to be based on things other than current scientific information. Often, the historical record continues to show, environmental issues are ventilated in the courts, and this body has become the debate forum where adversarial proceedings of environmental health are first addressed and eventually decided. This is unfortunate, because many of the environmentally-related health problems are extremely complex. Additionally, some are difficult, if not impossible, to study because of conflicting variables and elusive epidemiology. The control and prevention of environmental health hazards will be a perennial challenge to public health practitioners, and will doubtless taunt the imagination and ingenuity of scientists in the future (Frank, 1985).

The subject of veterinary drug residues in food covers a broad and diverse range of disciplines. It also has significant social and economic implications both at a national level and internationally, because actions taken by one country could impact – at times with serious consequences – on both agricultural trade and practices. Equally relevant is the prevailing interest of consumers who are demanding assurances that the drugs being used are safe, and will not result in unsafe residue levels in animal products (WHO, 1988).

In cases of contamination, however, people and animals became ill long before the responsible contaminant was identified. Often, no one was aware, or even suspected, that the offending substance was in the food.

Many of the major contamination incidents have followed this pattern, e.g. PCB, mercury, PBB. The many unpredictable and uncontrolled sequences associated with environmental contaminants, and the challenges they pose to public health, mandate that vigilance become a constant, and caution an absolute.

1.4 REFERENCES

Agricultural Science and Technology Council. 1981. "Antibiotics in Animal Feeds," *Report*, 88:1–79.

Beran, G. W. 1987. "Use of Drugs in Animals, an Epidemiological Perspective," in *Dollars and Sense. Symposium on Animal Drug Use*, G. E. Stefan, ed., Rockville, MD: Food and Drug Administration, pp. 3–27.

Bland, J., ed. 1987. "Food Safety," *World Health*, Geneva, Switzerland: The magazine of the World Health Organization, p. 32.

Chanlett, E. T. 1979. *Environmental Protection*. New York: McGraw Hill, 2:28–31.

Food Marketing Institute. 1988. "Trend: Consumer Attitudes and the Supermarket," Food Marketing Institute.

Frank, A. L. 1985. "The Status of Environmental Health," in *Public Health and Preventive Medicine*, J. M. Last, ed., Connecticut: Appleton-Century-Crofts, 12:495–497.

Guest, G. B., S. C. Fitzpatrick and L. M. Crawford. 1987. "Safety of Meat from Animals Treated with Naturally-Growing and Synthetic Hormones," presentation at *Journees Internationales Societe Francaise de Toxicologie*, Toulouse, France.

Kilara, A. and H. D. Graham. 1982. *The Safety of Food*. Westport, CT: AVI Publishing Company, Inc., pp. 528–532.

Marshall, E. 1980. "Antibiotics in the Barnyard," *Science*, 208:376–379.

Miller, M. A., ed. 1987. *FDA Veterinarian*, 11(111):1–2.

Murray, B. E. 1984. "Emergence of Diseases Caused by Bacteria Resistance to Antimicrobial Agents," in *CRC Handbook Series in Zoonoses*, Section D, Antibiotics, Sulfonamides and Public Health, J. H. Steele and G. W. Beran, eds., Boco Raton, FL, pp. 201–216.

NAS. 1980. "The Effects on Human Health of Subtherapeutic Use of Antimicrobials in Animal Feeds," Washington, D.C: National Research Council, pp. 1–21.

Nichols, D. R. and T. F. Keys. 1984. "An Historical Overview of Antibiotics and Sulfonamides in Society," in *Handbook Series in Zoonoses*, Section D, Antibiotics, Sulfonamides, and Public Health, G. W. Beran and J. H. Steele, eds., Boco Raton, FL, 1:35–43.

Scroggins, C. D. 1987. "Stepping Stones or Stumbling Blocks: A Consumer Perspective of Animal Drug Use," in *Dollars and Sense. Symposium on Animal Drug Use.* G. E. Stefan, ed., Rockville, MD: FDA, pp. 53–58.

Storclz, W. J. 1978. "C + EN's Top Fifty Chemical Products and Producers," *Chem. and Eng. News*, 56(18):33.

U. S. Congress, Office of Technology Assessment. 1988. *Pesticide Residues in Food.* Washington, D.C.: U.S. Government Printing Office, pp. 3–4.

U. S. Department of Agriculture, Food Safety Inspection Service. 1987. *Economic Impact of the European Community's Ban on Anabolic Implants.* Washington, D.C.: Food Safety and Inspection Service, pp. 528–532.

World Health Organization. 1988. "Evaluation of Certain Veterinary Drug Residues in Food," *Technical Report Series 763.* Geneva: WHO, pp. 7–29.

CHAPTER 2

Pharmacological Principles of the Disposition of Drugs and Other Xenobiotics

JUDI WEISSINGER, D.V.M., Ph.D.[1]
LESTER M. CRAWFORD, D.V.M., Ph.D.[2]

2.1 INTRODUCTION

THE quantitative aspects of absorption, distribution, biotransformation and excretion are very important in evaluating the therapeutic and toxic effects of drugs, in designing dosage regimes, in interspecies extrapolation, and safety and effectiveness considerations in administering drugs to animals. The intensity and duration of pharmacologic effect in two animals given the same dosage level may vary widely or be quite similar. The similarities and differences in effect are usually attributed to the level of drug attained at the site of action. The level of drug attained at the site of action is dependent upon the disposition kinetics of the drug, and the physiology of the species.

Pharmacokinetics provides a terminology for describing the way in which the body handles drugs. But pharmacokinetics does not exist alone, and pharmacokinetic determinations must be considered along with the extensive body of pharmacodynamic knowledge of classes of drugs, and the physiology of animals. The initial assumption inherent in all pharmacokinetic calculations is that the concentration of drug in the blood plasma at any time is proportional to, and representative of, the total concentration of drug in the body. This proportion is acceptable because it is based upon specific pharmacokinetic models which have been used to describe many of the processes involved in drug disposition. For any system where it can

[1]Director, Nonclinical Regulatory Affairs and Compliance, Research Triangle Park, North Carolina, USA.
[2]Executive Director, Association of American Veterinary Medical Colleges, Washington, D.C., USA.

be demonstrated that this initial assumption is not valid, the pharmacokinetic approach to describing the disposition of drugs is inapplicable.

Safety and effectiveness of a drug in an animal may be evaluated using pharmacokinetics. In order to achieve its effect, a drug must be administered to an animal in an appropriate dose. The site of administration must be such that the appropriate dose will be absorbed for transport to the site of action, and maintain that concentration for a suitable period of time. The intensity of the response will be related to the concentration at the site of action, which is, in turn, proportional to the blood level of the drug. If the concentration at the site of action is above a certain level, the animal may demonstrate a toxic response. Thus, with knowledge of the safety factor for drug response (therapeutic index), the plasma concentration of a drug from administration until it is eliminated from the body, plus the proportion of drug which is distributed to the tissues compared to the concentration in the plasma, the effectiveness of a drug and its safety in an animal can be assessed.

Absorption is described by determining the rate constant for absorption into the bloodstream at successive time intervals after administration. Distribution is described by determining the rate constant of the intravenously administered dose as it equilibrates to the body in general, correcting for any rapidly occurring metabolism and excretion. If distribution is determined after the same dose is administered intravenously, intramuscularly, orally, or via any other route, the fraction of drug absorbed by a given route may be determined. Biotransformation is described by determining the half-life of elimination, a rate constant describing elimination of the administered dose from the body in general.

All of these parameters may be determined from a semi-log graph representing the drug levels in the blood (ordinate) over the elimination time (abscissa) both before and after the drug in the plasma has achieved equilibrium with the total drug present and being eliminated from the body. The zero-time intercepts of the distribution phase and the elimination phase may be extrapolated graphically. With the additional knowledge of the rate constants of distribution and elimination and the fraction absorbed, additional parameters such as the apparent volume of distribution, volume of the central compartment, and systemic body clearance may be calculated, as well as the plasma concentration at any time after administration of a drug.

Bioequivalence of two drugs in one animal, or of the disposition of one drug in two different animals, may be demonstrated statistically by evaluating the equality of the data accumulated from measuring blood concentration-versus-time from shortly after the initial dose by any route of administration until 95% of the maximum concentration (four half-lives) has occurred.

Withdrawal time based upon residue levels predicted from pharmacokinetic models will be useful when the models developed are shown to accurately reflect and predict the tissue levels of a given drug. No extrapolations may be made until the basic experimentation is done, and these data evaluated for results and reproducibility. Then, a prediction may be made regarding the time it takes for a particular tissue to have drug residues below a given level.

For drugs that have measurable residues in tissues, but have levels in the blood below the limit of detection, equivalence has been considered on the basis of comparison of residue depletion curves. For drugs that have been shown to have no measurable absorption, recovery from the feces over time may adequately define the disposition profile.

2.2 ABSORPTION

A drug must be administered to an animal in appropriate concentrations to effect its action. The site of administration must be such that appropriate concentrations will be transported to the site of action, and will be maintained for a sufficient period of time, resulting in the drug's action. The intensity of response is related to the concentration of drug at the site of action. Residues may be measurable in tissue after some or all of the drug has been systemically absorbed. Absorption involves drugs passing into the bloodstream via capillary walls. Absorption is affected by physicochemical factors and the route of administration (Curry, 1977). The processes for drug absorption include passive transport, active transport, facilitated diffusion and pore transport. The physicochemical factors that affect drug absorption across the cell membrane involve properties of both the membrane and the drug (Weissinger, 1983a).

2.2.1 PASSIVE DIFFUSION

Passive diffusion—or simple diffusion—is the most common mechanism of absorption. The rate of transport of drug across the membrane is proportional to the the concentration gradient (always moving from an area of high to low concentration), the membrane thickness and the cross-sectional area of the membrane exposed to the drug solution. Passive diffusion is possible since the membrane is a phospolipid bilayer. Passive diffusion is prohibited where molecules are highly ionized and not lipid soluble. Predictors of the ability of a molecule to be absorbed via passive diffusion include the oil:water partition coefficient (*Pc)*, and the negative logarithm of the acidic dissociation constant (pK_a).

The PC is the ratio of the lipid solubility of a molecule to the aqueous

TABLE 2.1. Partition Coefficients of Selected Barbiturates.

Agent	Approximate Duration	Partition Coefficient	pK_a (weak acid)	% Absorbed
Thiopental	5 min	580	7.4	Higher
Secobarbital	25 min	52	7.9	
Pentobarbital	2 hr	39	8.0	
Phenobarbital	4–6 hr	3	7.3	Lower

Source: Weissinger, 1983b.

solubility of a molecule. It is usually calculated from in vitro experiments, where a given amount of drug is added to a test tube containing equal amounts of water and an oil such as octanol, chloroform, or olive oil. After shaking and incubating at 37°C, the concentration of drug in each phase is determined, and the ratio of the concentrations is the *Pc*. Obviously a *Pc* equal to 1 indicates equal solubility in both oil and water, greater than 1 equals more lipid solubility, and less than 1 indicates a drug is more aqueous soluble. This ratio predicts which drugs will not be readily absorbed passively, and may need to be introduced directly into the bloodstream. When comparing the *Pc*s of several drugs in a class, it is important that the oil used is consistent for all the drugs being compared. For example, the *Pc* of antipyrine in olive oil:water is 0.3 and in $CHCl_3$:water is 28. Absorption by passive diffusion tends to favor molecules with *Pc* > 1 and which exist primarily in the un-ionized state. Examples for barbiturates (pK_a 7.3–8.0) are listed in Table 2.1.

The second predictor of absorption via passive diffusion is the pK_a. The pK_a is defined as the negative logarithm of the acid dissociation constant of a drug. The Henderson-Hasselbach equation (see Figure 2.1) is used to calculate the degree of ionization of a drug at equilibrium (Baggot, 1977). For strong to weak acids, the pK_as range from <1 to 14, and for strong to weak bases, the pK_as range from 14 to <1.

a. For an acid

$$pK_a\text{-pH} = \log \frac{\text{unionized}}{\text{ionized}}$$

For a weak acid, $pK_a = 4$, a pH of 8, the unionized:ionized ratio would be $1/10^4$, or 99.99% ionized. For a weak acid with a $pK_a = 8$, at a pH of 6, the unionized:ionized ratio would be $10^2/1$, or 99% unionized.

b. For a base

$$pK_a\text{-pH} = \log \frac{\text{ionized}}{\text{unionized}}$$

For a weak base with a pK_a of 10, at a pH of 8, the ionized:unionized ratio would be $10^2/1$, of 99% would be ionized. For a weak base with a pK_a or 6, at a pH of 8, the ionized:unionized ratio would be $1/10^2$, or 99% would be unionized.

Figure 2.1 Henderson-Hasselbach equation.

TABLE 2.2. pK_a Values for Some Common Pharmacological Agents Used in Veterinary Medicine.

Acids	pK_a	Bases
Penicillins	2.5	Metronidazole
Aspirin, methylene blue	3.5	
Phenylbutazone	4.5	Amidropyrine
Sulfisoxisole	5.0	
Sulfadimethoxine	6.0	Histamine
Sulfadiazine	6.5	Trimethoprim
		Papaverine
Sulfamerizine, nitrofurantoin	7.0	Gentamicin
		Neomycin
Sulfamethazine, barbiturates	7.5	Lidocaine
Phenytoin		Dihydrostreptomycin
Sulfapyridine	8.0	Morphine
		Fentanyl
	8.5	Dilantin
		Erythromycin
	9.5	Propranolol
		Chlorpheniramine
	10.0	Antihistamines
Sulfanilamide	10.5	Quinacrine
Sulfaguanadine	12.0	

Source: Weissinger, 1983b

When the pK_a equals the pH of the medium, 50% of the drug is ionized and 50% is un-ionized. As the un-ionized portion crosses the membrane into the bloodstream, there is a continuous redistribution of the remaining ionized form to the un-ionized form so that eventually, all of the drug is absorbed by flux across the membrane. Table 2.2 lists the pK_a values for some common pharmacological agents used in veterinary medicine.

2.2.2 ACTIVE TRANSPORT

Active transport is different from passive transport because it:

(1) Proceeds against a concentration gradient (moves from low to high concentration)
(2) Is carrier mediated
(3) Requires energy

Energy is usually in the form of adenosine triphosphate (ATP). This type of absorption is site-specific because it only occurs where the carrier molecules are present. Carriers are also specific for certain drugs, so not all drugs may be transported by this process. Since a finite number of carrier molecules may be involved, saturation may end the process. Glucose, ionized acids, quaternary ammoniums and sulfonic acid ions are examples of compounds that are carried by active transport against a concentration gradient. Competitive and noncompetitive inhibition have been demonstrated, which is due to the fact that carrier molecules are both specific and finite.

2.2.3 FACILITATED DIFFUSION

Facilitated diffusion is carrier mediated, but not energy dependent. This absorption mechanism acts along a concentration gradient as described for passive diffusion, but is carrier mediated like the active transport system. No ATP is required. However, absorption is saturable, site specific, and subject to competitive and noncompetitive inhibition.

2.2.4 PORE TRANSPORT

Pore transport is usually the mechanism of absorption reserved for small molecules. These small molecules, with molecular weights less than 300 daltons, are transported via diffusion through aqueous-filled channels in membranes.

2.3 ROUTE OF ADMINISTRATION

Drugs must be introduced into the body to achieve systemic effects. Introduction may occur by parenteral administration or enteral administration. Direct application to the desired site of action is used to achieve primarily local effects. The factors affecting absorption from the various routes of administration, and the advantages and disadvantages of each route, are discussed in the following paragraphs.

Parental administration is utilized to bypass the alimentary tract. Systematic effects may be achieved after parenteral administration by injection, inhalation and percutaneous. Drugs may be injected intravenously, intra-arterially (rarely used except diagnostically and for cancer chemotherapy), intramuscularly, subcutaneously, intraperitoneally, and intrapleurally.

Intravenous injections are often desirable because entry into the system depends only upon the rate of injection and not on absorption into the bloodstream. The advantages of intravenous injection are:

- Drug gets into the system with a minimum of delay.
- Large quantities of fluid may be administered over time.

- Careful regulation of blood concentration is possible.
- Poorly absorbed substances or very irritating drugs may be administered.

Although the advantages are substantial, intravenous administration is only amendable to water soluble substances. Disadvantages of intravenous injection include overdoses, and hypersensitivity reactions are more severe and the risk of embolism and infection is greater.

Absorption of drugs after intramuscular injection is influenced by several factors. The rate-limiting step for absorption is the rate of blood flow in the muscle. No cell in the body is >50 microns from a capillary, and adequate circulation is needed to maintain the concentration gradient that is so critical to passive absorption or facilitated diffusion into the bloodstream. The higher the vascularity of the tissue the better the absorption, because diffusion distances are decreased. The nature of the vehicle and the lipid solubility of the drug, where favorable, will potentiate absorption. Generally, the larger the molecule the longer the dissolution time, and consequently increased absorption time. The volume injected and the concentration of the drug in solution may have varying effects upon absorption, depending upon the potential for the drug molecule to come into contact with the absorptive surface. The advantages of administration by intramuscular injection are that the muscle can act as a depot, and the rate of disappearance of drug from the site of injection can be calculated. The main disadvantage, ignoring the discomfort to the animal, is that the volumes administered by this route are limited, compared to quantities used in intravenous administration.

Absorption after subcutaneous administration is influenced by the same factors as intramuscular administration (Booth and McDonald, 1982). Subcutaneous preparations may include insoluble suspensions, or solid pellets used as implants. However, this is a less than optimal route of administration for irritating substances which cause local necrosis.

Absorption after intraperitoneal administration is affected by the same factors affecting absorption after intramuscular and subcutaneous injection. Intraperitoneal injections are generally used to achieve rapid and complete absorption via the portal system in experimental animals, range animals, and swine. They are also used for peritoneal washes. This route of administration is not optimal where the drug administered is subject to *first pass* clearance by the liver (see section on biotransformation). The major risk associated with intraperitoneal administration is aseptic peritonitis.

Inhalational, intranasal, and intratracheal administration are normally reserved for vapors, aerosols, and gases (including anesthetics). Absorption is facilitated by smaller-sized particles, high lipid solubility ($Pc > 1$), sufficient pulmonary blood flow and a large, intact, absorptive surface area, as is present in healthy lungs. Administration by these routes is very rapid when several of the factors favoring increased absorption are combined. Administration may be used for both systemic and local application, simultaneously. The main disadvantage of respiratory tract administration is the susceptibility of the epithelium to irritation.

To achieve systematic effects with percutaneous administration, the drugs must pass through the stratum corneum, epidermal cells, sweat glands, sebaceous glands, or hair follicles before being absorbed. Only very lipid-soluble drugs (including organophosphates, insecticides, and solvents) will pass through the intact skin. The primary advantage of transdermal delivery is that it is a noninvasive technique. However, it is often difficult to obtain sufficient concentrations of the drug in the plasma or site of action to obtain the desired effect. Even when significant absorption has been demonstrated, it is often not reproducible because of intraspecies variability in transdermal absorption. During administration, if the skin is not thoroughly clean, lipid soluble toxicants, if present, may pass readily through the skin in sufficient amounts to produce toxicity.

Enteral administration is utilized for drugs that may be absorbed via the intestinal tract. Oral and rectal administration are the usual routes. Pills, tablets, capsules, elixirs, suspensions, and solutions may be administered per os – rectal suppositories and enemas. Absorption via this route is highly species specific and is affected by many factors. The presence of food in the stomach may reduce, increase, delay or have no affect upon absorption, depending upon the drug formulation. Adequate motility is necessary to transport the drug to the area of optimal absorption. The partition coefficients of the drugs are important in facilitating membrane transport, as is the pH of the gastrointestinal fluid in determining the percent of drug in the un-ionized form favoring absorption. The method of transport will be determined, in part by the size of the molecule. The rate of splanchnic blood flow will enhance drug absorption, and is influenced by the meal, e.g. a high-protein liquid meal may increase the flow, and a liquid glucose meal may decrease the flow. As with injections, the volume administered may have a varying effect depending upon the effect on the ability of the molecule to come into contact with the absorptive surface. Enteric coating of preparations will inhibit absorption in the stomach of carnivores and omnivores, and allow for intestinal absorption.

The rate of gastric emptying may increase or decrease absorption depending upon the site of absorption and several other factors. Physiological factors affecting absorption include posture of the animal, osmotic pressure, gastrointestinal contents and distention. Pathological factors affecting absorption include ulcers, obstructions, liver disease, trauma, and gastroenterostomies. Pharmacologic factors affecting absorption include products altering pH (e.g. antacids) anticholinesterases, anticholinergic drugs, narcotics, and products affecting acid secretion (e.g. autocoids).

Despite all the variability in absorption related to the factors affecting absorption, there are advantages to enteral administration. These include the relative safety and ease of reversibility, economic aspects relating to convenience of client administration and lack of necessity for sterility. Still, vomiting may result if irritating substances are administered; complexing and degradation of the drug may occur in the gastrointestinal tract; and it takes longer than it does after intravenous administration to reach therapeutic levels. Drugs that are rapidly metabolized by the liver may be inactivated before sufficient quantities reach the system, and some drugs are locally toxic to the intestinal epithelium. Ruminants have a high microbial intestinal population that may adversely affect the intact drug prior to absorption.

Where oral administration is preempted by vomiting, or when a drug is highly susceptible to first pass clearance by the liver, the rectal route of administration may be advantageous. Suppositories may be used in situations where vomiting precludes administration, or in states of unconsciousness. Drugs administered rectally do not have to pass through the liver prior to entry into the systemic circulation. Drugs irritating to rectal mucosa should not be considered for administration via this route. Because absorption via the rectum is often incomplete and irregular, this route should be reserved for products with wide therapeutic windows.

When local effects are desired in the integument, the synovium, the central nervous system, the mammary glands, or the uterus, then intradermal or topical, intra-articular, intrathecal, intramammary, or intrauterine administration may be most appropriate. For drugs used to treat the intestinal mucosa, enteral administration may be appropriate provided the drugs are not absorbed. The advantage of local administration is the ability to achieve a high concentration of drug at the immediate site of action, but it is necessary to keep in mind that all of these routes, depending upon the drug preparation, have the potential for facilitating inadvertent systemic exposure.

2.4 DISTRIBUTION

Once administered, the drug must be distributed to the site of action. After the animal has absorbed the drug into the bloodstream, it can be dispatched to the body fluid compartments. The distribution is initially governed by the lipid solubility of the drug, the cardiac output, and the regional blood flow at the sites of administration and action. Tissues at the site of action may attain a higher or lower concentration of drug than the plasma due to pH gradients, nonspecific binding, transport processes, and lipid solubility. Drug accumulation in body tissues may serve as a reservoir to delay onset or prolong action of the drug.

Distribution of drugs to the tissues and the equilibration of drug between blood tissue are affected by regional blood flow. The liver, kidney, heart, and brain are the most highly perfused tissues in the body, and are readily exposed to drugs. Drugs may exist as a reservoir in the blood by binding to plasma proteins, thus restricting distribution. The unbound fraction is the active fraction.

Most drugs that do bind are bound to a fixed degree by reversible bonds. The reversibility of the bonds follows the law of mass action where, as the concentration of unbound, un-ionized, drug leaves the bloodstream, a proportion of the drug bound to the plasma proteins becomes unbound to reestablish the equilibrium describing the fixed percent that is not bound. Binding of drugs to plasma proteins is restricted to the intravascular spaces. Albumin is the principle drug-binding plasma protein. It has multiple sites with differing affinities. Most sites show more extensive binding to weak acids than to weak bases. A decrease in total albumin levels will generally increase the absolute total amount of drug present that is unbound and therefore available to leave the bloodstream, but the percent of drug remaining that is protein-bound will often be constant.

Protein binding is affected by drug concentration and protein concentration. Protein-binding receptors are saturable, and, in overdose situations, a much higher percentage of drug than normal is available to leave the bloodstream. Competitive inhibition for albumin protein binding sites has been demonstrated (Goodman and Gilman, 1985). This competition extends to endogenous substances such as calcium and hormones. In addition to competing drugs and variable albumin concentrations, the pH, temperature, divalent cations, uremia, catabolic states and species variability may alter the degree of protein binding.

For drugs that are bound less than 80% (i.e., >20% of the drug is unbound), an alteration in protein binding of 5% will not significantly affect tissue concentrations of the free drug. However, when drugs that are bound greater than 95%, a 5% decrease in binding will result in the doubling of the unbound fraction; the alteration in tissue concentration of free drug is significant. The unbound fraction is considered to be active, and available to leave the bloodstream to distribute to the site of action.

Unbound, un-ionized, drugs cross the blood vessel membranes as described in the section on absorption (i.e., by passive diffusion, active transport, facilitative diffusion and pore transport). Entry of a drug into the cells is generally restricted to lipid soluble (un-ionized) drugs and small ions. Water soluble drugs (usually metabolites and ionized forms) will stay in the bloodstream and be transported to the kidneys for excretion, instead of being distributed to the tissues. The pH of the receiving tissue affects absorption much the same as the pH of the absorptive surface affects transport; the difference between the pH in the blood and the milk, urine, saliva, etc., affects the relative concentration of drug in these fluids.

Drugs binding to tissues may produce areas of high drug concentration, and may bind to cellular membranes and soluble or insoluble intracellular components. This type of binding is usually reversible. Depot binding results in a reservoir or storage, such as has been demonstrated in bone and fat. If the plasma concentration of a drug decreases, the drug may enter the plasma from these storage sites via the concentration gradient. Specific tissue receptors are often the binding sites for drugs, even though they are not the receptors. For example, after the plasma concentration for digoxin or quinine drops, the heart and liver still maintain high concentrations (Goodman and Gilman, 1985). In the kidney, aminoglycosides bound to proximal tubular cellular components remain for months after plasma concentrations have become undetectable.

An example of the effects of binding and degree of ionization on distribution of a weakly acidic drug (pK_a-2) that is 50% protein-bound are illustrated by the following: When the drug is administered orally to a sow (a stomach pH of 1) 91% will be un-ionized, and thus readily absorbed. In contrast, in the bloodstream, where the pH is 7, this drug will be 99.999% ionized, and will not readily pass through the blood vessels into the tissues. If, on the other hand, the pK_a of a drug is 6, the drug is virtually 100% un-ionized in the stomach and 9% un-ionized in the blood, allowing for higher amounts to distribute to the tissues. If the sow is lactating, and the pH of the milk is elevated because of an infection, only 1% in the milk is un-ionized, and as the drug crosses into the milk, it may become iontrapped, favoring accumulation in the milk. This principle can be useful for selecting antimicrobial compounds for systemic administration to treat mammary infections.

2.5 BIOTRANSFORMATION

Some drugs are eliminated from the body unchanged, but most drugs undergo biotransformation. Biotransformation refers to the process by which drugs are altered by interaction with substances present in the body fluids i.e., the biologic fate of drugs. Biotransformation often enables a compound to be converted to a form that is readily excreted by the liver and kidney. Sometimes biotransformation enables a drug to be converted to an active form. However, these conversions almost always produce a metabolite that is more polar (aqueous soluble) than the parent compound. The liver is the most important site of drug biotransformation. Other tissues capable of biotransformation include the kidney, lung, brain, adrenals, skin, blood, neurons, and the gastrointestinal tract. Biotransformed drugs delivered to the intestinal tract may, additionally, be returned to the parent drug form via enzymes from intestinal bacteria such as beta-glucuronidase (Booth and McDonald, 1982).

Biotransformation may produce metabolites by synthetic or nonsynthetic mechanisms. Nonsynthetic biotransformation, including oxidation, reduction and hydrolysis, produces a change in activity of the drug via a change in chemical structure without increasing the molecular size of the molecule. *Oxidation,* facilitated by oxidases, describes the process whereby a drug molecule is combined with an oxygen group or released from a hydrogen group. *Reduction,* facilitated by reductases, describes the process whereby a drug molecule is combined with a hydrogen group or released from an oxygen group. *Hydrolysis,* facilitated by esterases and amidases, describes the process whereby a drug molecule's chemical bonds (e.g., esters and amides) are split by addition of water. These processes, while normally resulting in inactive metabolites, may also result in active metabolites of the drug.

Synthetic biotransformations produce a drug molecule which is enlarged by combining it with inactivating substances (conjugation). Inactivation of the parent drug almost always occurs when the conjugations occur at an OH, COOH, NH_2 or SH. Glucuronides are synthesized using UGDPA and transferases in hepatocyte liver microsomal enzymes by using endogenous glucuronic acid for conjugation. Acetylation, the main pathway for biotransformation of sulfonamides, occurs in the reticuloendothelial cells of the liver. Acetylation with glycine and glutamine is a mechanism of biotransformation resulting in inactivation in most mammalian species, but in poultry such as ducks, geese, hens, and turkeys, ornithine is the amino acid used for conjugation. Sulfate conjugation is facilitated by sulfurases using endogenous thiosulfates sulfate groups, and *O-*, *S-*, and *N*-methylation via methyltransferases using endogenous methionine as a methyl donor.

Most biological oxidations, glucuronide conjugations and some hydroxylations and reductions are catalyzed by enzymes predominatly synthesized by the smooth endoplasmic reticulum (SER) of the liver. Examples of some common animal drugs are listed in Table 2.3. These microsomal biotransformations require NADPH and oxygen. Drugs and endogenous steroids must be lipid soluble and un-ionized at physiological pH to get to the SER.

Drugs that inhibit microsomal activity will increase the duration of action of that drug (Table 2.4). This delayed biotransformation may be seen within minutes of administration of a drug such as chloramphenicol or carbon monoxide. These enzymes are saturable, which means that an inhibi-

TABLE 2.3. Food Animal Drugs Known to Be
Biotransformed to Varying Degrees
by Liver Microsomal Enzymes.

Chlorpromazine
Dexamethasone
Diphenylhydantoin
Disophenol
Hormones
Neostigmine
Pentobarbital
Sulfonamides
Vitamin K

Source: Weissinger, 1983b.

tor may compete for biotransformation with another drug that was also biotransformed by microsomal enzymes. Stimulation (induction) of microsomal enzyme synthesis, and thus stimulation of the rate of biotransformation of drugs (contrary to inhibition of activity), is delayed, while increased proliferation of the endoplasmic reticulum occurs (Table 2.5). Induction of microsomal enzymes usually occurs over 7–10 days. The resultant increase in metabolic ability may enhance the biotransformation of endogenous compounds such as bilirubin, steroid hormones and vitamin D. Where the natural compensatory feedback mechanisms for increasing production of these endogenous substances is working, induction presents no impediment to a single drug's biotransformation. In fact, the main drawback of induction is the increased biotransformation of the drug itself, thus requiring increased doses to achieve the same effect, producing an apparent tolerance to the drug.

Some oxidation, reductions and hydroxylations occur in other tissues. Neuronal monamine oxidase is useful for biotransformation of catecholamines, xanthine oxidase methylxanthines, and plasma pseudocholinesterase for cholinergic compounds. Intestinal bacterial enzymes, rumen

TABLE 2.4. Chemicals That Inhibit the
Activity of Liver Microsomal Enzymes.

Carbon monoxide
Carbon tetrachloride
Chloramphenicol, metapyrone
Parathion, malathion
Quinidine

Source: Weissinger, 1983b.

TABLE 2.5. Drugs That Are Known to Induce
Liver Microsomal Enzyme Synthesis.

Chlorpromazine
DDT
Diazepam
Diphenhydramine
Diphenylhydantoin
Estradiol
Griseofulvin
Hexobarbital
Methoxyfluorane
o, p'–ddd
Phenobarbital
Phenybutazone
Protestogens
Androgens

Source: Weissinger, 1983b.

hydrolytic and reduction microfloral enzymes play a major role in the metabolism of enterically administered or excreted drugs.

Biotransformation is affected by several factors, including age, species differences, hormones, diet, disease, environment, and foreign compounds. The activity of the liver microsomal enzymes is low in newborns and aging animals, resulting in a slower rate of biotransformation. Species differences in dose and response may often be attributed to biotransformation differences. For example, cats have comparatively low levels of glucuronyl transferase, and are not able to inactivate drugs by this mechanism, although they do conjugate endogenous substances. Swine demonstrate very little capacity for sulfate conjugation, and dogs cannot acetylate primary amines such as sulfonamides. Where alternate pathways for biotransformation exist, or where drugs are excreted primarily unchanged, toxicity may be avoided.

Differences in biotransformation capability may affect the dose, duration, and toxicity. Table 2.6 lists a species comparison of the half-lives of elimination of phenylbutazone, a liver microsomal enzyme inducer with an active metabolite, oxyphenbutazone.

Hormones such as sex steroids and thyroxine have been shown to stimulate or inhibit biotransformation of exogenous compounds. Foreign compounds such as other drugs, insecticides and food additives have been shown to affect biotransformation. Inadequate protein intake, approaching starvation, may decrease the rate of biotransformation. Diseases of the liver may also interfere with normal biotransformation capacity. Increases in biotransformation occur when fever increases the metabolic rate.

Dosage regimes, withdrawal times, and milk withholding times are all based upon administration of drugs to healthy animals. They are determined where absorption, distribution, metabolism, and excretion are consistent with normal states of health. It is important to recognize that many factors influence the biotransformation of drugs, and thus the half-life of elimination from both the body and the edible tissues. Adjustments in dosage, duration, withdrawal time and milk withholding times should always be considered where biotransformation of drugs may be affected by existing conditions.

2.6 EXCRETION

Drugs may be eliminated from the body by several vehicles including urine, bile, exhaled air, milk, saliva, and sweat. Drugs must be removed from the body fluids before elimination occurs. The most important route of excretion, and thereby termination of drug action, for most drugs is the kidney. Extrarenal sites of excretion include the liver, lung, mammary gland, sweat gland, salivary glands and intestinal mucosa. Excretion therefore refers to the removal of drugs from body fluids by the kidney, liver, lung, and other organs.

Renal excretion of drugs is facilitated by kidney filtration at approximately one-tenth of the cardiac output per minute. Excretion of drugs via the kidney occurs primarily by glomerular filtration, but also by proximal tubular secretion and distal tubular absorption. Glomerular filtration relies on the size of the pore. The glomerulus apparatus makes no distinction be-

TABLE 2.6. Comparison of the Half-Life of Phenylbutayane in Various Species.

Man	72 hours
Ox	55 hours; 42 hours*
Goat	19 hours (male), 15 hours (female)
Cat	>18 hours
Rat	6 hours
Dog	6 hours
Baboon	5 hours
Swine	4 hours
Horse	3.5 hours
Rabbit	3 hours

* Both values have been reported with no gender associated, and it is assumed that there is a biphasic curve suggesting two distinct half-lives (probably a factor of conversion to oxyphenbutazone, the active metabolite).
Source: Weissinger, 1983b.

tween the ionized and un-ionized fractions of drug. The rate of glomerular filtration varies among species, and the driving force is the blood pressure (Table 2.7). Drugs that decrease the cardiac output or the total peripheral resistance also decrease the glomerular filtration rate. Even though there is no distinction between ionization states, protein binding precludes excretion via glomerular filtration. Only nonprotein bound drugs are filtered. This exclusion is probably a factor of molecular size, as molecules with molecular weights greater than 50,000 daltons are not freely filtered at the glomerulus.

Proximal tubular secretion is an energy-dependent active-transport mechanism. Specific high-affinity carrier proteins in the proximal tubule transport drugs into the tubule for elimination via the urine. Proteins also remove organic acid and bases from plasma protein binding sites and transport these drugs into the tubule. Because it is a carrier-mediated system, it is a saturable system, and drugs may also compete for transport where similar carriers are used. This principle of competition for excretion may be used where it is desirable to prolong the level of drug in the body, or, alternatively, where a preferential excretion is demonstrated.

Distal tubular resorption occurs with many lipid soluble drugs, and drugs which are un-ionized. Reabsorption in the distal tubule may be via active transport with saturation and competition, or may be via passive diffusion. Water is reabsorbed, thus producing a high concentration in the urine and a concentration gradient favoring drug resorption. The standard determinants of passive diffusion apply to drug reabsorption, including pK_a of the drug, pH of the urine, and rate of blood flow. Where it is desirable to increase the duration of drug in the body, manipulating the pH of the urine to favor the un-ionized form of the drug may significantly increase the half-life of elimination. In overdose or toxicity situations, where it is desirable to increase excretion of a drug, manipulation of the pH of the

TABLE 2.7. Glomerular Filtration Rates for Various Species.

Species	GFR (mL/kg/min)
Rat	10
Dog	4.0
Goat	2.2
Sheep	2.0
Sheep	1.8
Human	1.8
Horse	1.7
Cat	3.5

Source: Weissinger, 1983b.

TABLE 2.8. Drugs Which Require Modification of Doses in Renal Insufficiency.

Atropine	Neostigmine	Digoxin
Procainamide	Mannitol	Mercurials
Acetazolamide	Chlorthiazide	Spironolactone
Barbital	Acetylsalicilic acid	Colistin
Polymyxin	Kanamycin	Streptomycin
Gentamicin	Vancomycin	Neomycin
Penicillins	Cephalothins	Erythromycin
Lincomycin	Clindamycin	Sulfonamides
Tetracyclines	Nitrofurantoin	Methotrexate
Iodine	Chelating agents	Methenamine

urine to favor the ionized form of the drug may decrease reabsorption. Biotransformation produces metabolites which are generally more polar than the parent compound, and thus favors excretion.

Renal insufficiency significantly affects drug excretion, even at one-third renal function. A reduced dose of a drug may be necessary because plasma levels could become high, or an increased dosing interval for administration of a drug may be necessary as the rate of excretion is inhibited (Tables 2.8 and 2.9). Switching to a drug that is not primarily dependent upon renal excretion will aid in avoiding dosage and interval adjustments, and buildup of residues.

While most compounds are excreted primarily by renal mechanisms, some drugs are partially or completely excreted through the bile. The liver is capable of transporting a variety of drugs via passive and active secretion into the bile using a transport system. Drugs excreted in the bile are usually already biotransformed in the liver, possess a high degree of polarity, and are usually conjugated. Certain drugs may compete with each other at the transport system based upon their acidic, basic, or neutral properties. As with all systems involved in drug disposition, consideration should be given to the increased tissue accumulation that may occur with hepatic disease and parasites.

There is extensive species variation in the general ability to secrete drugs in the bile. Rats, dogs, and chickens are characterized as good

TABLE 2.9. Drugs Which Do Not Require Modification of Doses in Renal Insufficiency.

Propranolol	Doxycycline	Quinidine
Diazepam	Phenytoin	Narcotic analgesics
Lidocaine	Penothiazines	Novobiocin
Procaine	Short acting barbiturates	Chloramphenicol

biliary excreters. Poor biliary excreters include rabbits, guinea pigs, monkeys, and humans. Cats and sheep fall between these two groups. When drugs are excreted via the bile, they may either be eliminated via the feces, or they may be hydrolyzed to the parent compound by bacterial enzymes and reabsorbed via enterohepatic circulation.

Salivary secretion is important in species where reabsorption of drugs from the gastrointestinal tract may increase the half-life of elimination. Exhalation of ketones, anesthetic gasses and products of drug metabolism, such as carbon dioxide and water, also may account for excretion of drugs. Acidity of milk favors mammary excretion of basic drugs, but doesn't exclude acidic drugs. In cases of mastitis where the pH of the milk is less acidic, both the ionized fraction in the milk and the excretion via the milk are altered. Considerations as to changes that will increase milk withholding time should be included in the determination for marketing milk after treating mastitis. These considerations should also include colostral milk after prior treatment of dry-cow mastitis.

2.7 BIOAVAILABILITY

Bioavailability is the term used to indicate a measurement of both the relative amount of an administered drug that reaches the general circulation, and the rate at which this occurs (Goodman and Gilman, 1985). The general assumption is usually made that as long as the amount of active ingredient is constant, no differences would be caused by variation in the pharmaceutical agents studied. A determination of equal bioavailability of two formulations of the same drug, both containing equal amounts of active drug (bioequivalence), is made by determining if the blood plasma levels are equivalent in the same time period after administration.

The entire curve describing the concentration of the drug in the blood versus time is referred to as the bioavailability curve of the drug. This curve may be characterized by three parameters—the time necessary for the drug administered to reach its maximum concentration, the actual maximum concentration of the drug, and the area under the statistical curve which describes the absorption, distribution, biotransformation, and excretion of the drug.

In determining if two drugs are bioequivalent (have equal bioavailability), it is not enough to show just that they achieve the same ultimate concentration in the blood, nor is it enough to show that they reach their maximum concentration at the same time. Similarly it is not enough to just demonstrate that the areas under the bioavailability curves of the two drugs are equal. Bioequivalence may only be demonstrated by proving that the bioavailability curves for two drugs are virtually superimposable with respect to peak, time to peak, and area.

There are several formulation factors which affect bioavailability. A drug will dissolve more rapidly if the surface area in contact with the solution is increased. It will be absorbed more rapidly if the particles in contact with the absorptive surface are increased. As some of the drug is absorbed, more particles may separate from the main concentration of drug and move into the diffusion layer. Changes in hydration state can affect aqueous solubility, and changes in polymorphic forms can alter the solubility.

Variations in bioavailability often occur because, in addition to the active constituent, it is generally necessary to include one or more of the following inert ingredients:

- dilutents to increase bulk
- granulating/binding agents to aid particle aggregation
- lubricants to prevent adherence to machinery
- disintegrating agents to effect rapid disintegration
- antioxidants to prevent decomposition
- potassium carbonate to control pH in moist surroundings
- dyes to color tablets
- coating to prevent tablet dissolution in stomach
- special formulations for sustained release

Variations in bioavailability may be demonstrated in different species. A product with a known bioavailability in one species will have a bioavailability in a different species dependent upon the species differences in absorption, distribution, biotransformation, and excretion. Withdrawal periods determined for invermectin after injection in ruminants (cattle, goats, buffalo, and reindeer) will differ greatly because the bioavailability and, ultimately, the tissue concentrations (especially the fat accumulation and concentration) are not consistent across species. Oral administration of a given product to different species usually demonstrates differences in absorption based upon enteric differences in pH, flora, and gastric emptying time. It is undesirable to generalize elimination, residue depletion, and milk withholding times across species, because the amount of drug that reaches the circulation, and the rate at which this occurs, i.e., bioavailability, is inconsistent.

2.8 RESIDUE CONSIDERATIONS

Computer simulation techniques are used to determine withdrawal periods and monitoring for residues based upon dose, species, health state, and known pharmacokinetics. The use of computers is also preferred to generate dose scaling across species based upon the significant physiologic and pharmacologic parameters involved in disposition of the drug, and for drug administration regimes with respect to effective doses. With the increasing knowledge relating the qualitative and quantitative aspects of drug metabolizing enzymes to the metabolites produced in each species, comparative metabolic profiles may be predictable at some time in the future. The ultimate regulatory goal is to provide a safe product to the consuming public. The validity of alternative approaches will come with experience, research to develop an adequate data base, and an increased understanding of the relationship of the use of drugs to animal husbandry.

2.9 REFERENCES

Baggot, J. D. 1977. *Principles of Drug Disposition in Domestic Animals.* Philadelphia, PA: W. B. Saunders Co.

Booth, N. H. and L. E. McDonald, eds. 1982. *Veterinary Pharmacology and Therapeutics.* Ames, IA: Iowa State University Press.

Curry, S. H. 1977. *Drug Disposition and Pharmacokinetics.* Oxford: Blackwell Scientific Publications.

Goodman, L. S. and A. Gilman. 1985. *The Pharmacologic Basis of Therapeutics.* New York, NY: MacMillan Publishing Company.

Weissinger, J. 1983a. "Pharmacological Principles of Drug Interactions," *Compendium on Continuing Education for the Practicing Veterinarian*, 5:839–850.

Weissinger, J. 1983b. Course Syllabus, Agents of Disease III—Veterinary Pharmacology and Toxicology, Colorado State University, Chapters 1–7.

CHAPTER 3

Principles and Implementation of Residue Programs in Meat and Poultry Inspection

RONALD E. ENGEL, D.V.M., Ph.D.[1]

3.1 INTRODUCTION

THE past four decades have seen a remarkable scientific revolution in agriculture, in which chemical technology has played a crucial role. Chemicals from a variety of sources may enter the food supply, including some that may have short- or long-term harmful effects in humans. The exposure of animals to environmental contaminants, or the use of pesticides or animal drugs in ways that do not conform to approved uses, can leave unacceptable levels of chemical residues in edible tissues. National health authorities have responded to the problem of residues by enacting laws regulating the use of pesticides and drugs in agriculture, and by limiting the exposure of the human population and its food supply to industrial and environmental contaminants. Programs have been developed to assess the probable and acceptable human intakes of potentially harmful substances in food. These programs encourage the concept of good agricultural practices (GAP) at the farm level.

Recognizing the importance of harmonious and positive approaches to the prevention of unsafe levels of residues in animal food products, seven residue control principles have been developed. They consist of:

(1) The establishment of minimum residue levels or tolerances (MRL's) in accordance with GAP

(2) Validation of analytical methods for drug and pesticide residues

(3) Laboratory quality assurance programs (The performance and ex-

[1]Assistant to the Administrator for International Scientific Liaison, Food Safety and Inspection Service, USDA, Washington, D.C. USA.

perience of analysts must be rated on their ability to detect and quantify compounds of interest at minimum detectable limits.)

(4) Criteria for compound evaluation and selection for monitoring

(5) Appropriate statistical design criteria for a residue monitoring program (Sampling and regulatory procedures should be followed which are compatible with the kind of residue data needed from monitoring and for enforcement programs, and which maximize use of finite resources.)

(6) Education and residue avoidance programs

(7) Enforcement programs

These principles form a solid and broad foundation for the establishment of a national system of residue control. The principles have been applied with considerable success in the U.S.

Since 1967, the Food Safety and Inspection (FSIS) of USDA has conducted a program by applying these principles to help prevent the marketing of animals containing illegal residues of drugs and other chemicals. FSIS is responsible for enforcing the Federal Meat Inspection Act and the Poultry Products Inspection Act. Under these laws, the agency ensures that meat and poultry products in interstate and foreign commerce are wholesome, not adulterated, and properly marked, labeled, and packaged. To accomplish this, about 7,000 federal inspectors and veterinarians carry out inspection in some 7,200 meat and poultry plants throughout the country. This is the largest inspection force in the federal government, both in absolute numbers and in the ratio of inspectors to regulated facilities. Each year, over 2 million laboratory analyses and inplant tests are performed on over 450,000 samples to support our residue and microbiological monitoring and surveillance programs. Each chemical residue sample costs an average of $140 to process—including collection, shipping, testing, cost for laboratories and scientific support. In 1991, USDA inspected 120 million head of livestock, 5 billion birds, and 127 billion pounds of processed products. FSIS carries out its residue control responsibilities in cooperation with other federal agencies. For the most part, these efforts are coordinated among FSIS, the Food and Drug Administration (FDA) and the Environmental Protection Agency (EPA).

3.2 ESTABLISHMENT OF MRL's (TOLERANCE OR ACTION LEVELS)

Through this interagency coordination, the first of the seven principles of residue control, and the establishment of acceptable residue levels are put into practice. In their respective areas of responsibility, EPA and FDA establish the tolerances for compounds to which livestock and poultry are exposed. They also determine the approved methods of use of compounds on specific crops or animals to ensure that tolerances will not be exceeded. EPA establishes tolerances for pesticides, while FDA sets tolerances for animal drugs and other nonpesticide contaminants in food. Where formal tolerances are established, FDA and EPA, as appropriate, recommend action levels to FSIS upon request for unavoidable contaminants.

Under the provisions of the Food, Drug and Cosmetic Act (FDCA), FDA has authority over the approval and regulation of all human and animal drugs. The approval procedure for these drugs involves, among other things, consideration of the efficacy of the compound proposed for use, safety to the target species, human food safety, and methods of analysis. When animal drugs are approved, and use restrictions, withdrawal periods, and tolerances are established, FDA approves medicated animal feeds (U.S. Code, Title 21). The Environmental Protection Agency (EPA) enforces the Federal Insecticide, Fungicide, and Rodenticide Act (FIFRA). Under the law, the agency has the responsibility for regulation and premarketing registration of pesticides, including the establishment of tolerances, or granting tolerance exemptions (U.S. Code, Title 40). EPA is also responsible for the regulation of industrial chemicals under the Toxic Substances Control Act (TOSCA) and for the disposal of industrial wastes under the Resource Conservation and Recovery Act (RCRA). In enforcing residue limits in meat and poultry products, FSIS relies on information and tolerances established by FDA and EPA.

3.3 ANALYTICAL METHODS

The FSIS residue program uses recognized methods – both official and unofficial – for the analysis of samples from slaughtering establishments. Official methods include methods approved by the Association of Official Analytical Chemists (AOAC), validated methods, Federal Register methods, and historical official methods. AOAC methods are those that have been subjected to extensive interlaboratory study involving at least five laboratories before official acceptance by AOAC (AOAC, 1984). Validated methods have been subjected to interlaboratory study by three laboratories and reviewed by a peer group of government scientists. Federal Register methods are those that have been published in the *Federal Register* and later incorporated into the Code of Federal Regulations. Historical official methods are those that have been considered to be the best available at the time of initial acceptance, and continued in use over an extended period of time in the absence of a more effective method.

Unofficial methods are those requiring additional product analysis before regulatory action is taken. These methods are used to determine the need for official methods in product testing, and as a preliminary phase of official methods development. There are three types of unofficial methods – the pilot study, the screening, and the published method.

The pilot study method is developed in a single laboratory, where statistics on methods performance are generated. The method is used to determine the need for additional testing or the development of official methods. The screening method is semiquantitative and is used to determine the presence or absence of compounds. A screening method permits rapid processing of large numbers of samples in less time than would be required by an official method. Positive results at or above specific levels require further analysis by an official method. A published method is one that has been published and subjected to a ruggedness test in an FSIS laboratory, but has not been thoroughly evaluated outside the originating laboratory. It may be used in nonrecurring analyses, and requires the development of a rigorous protocol for sample analysis. Before acceptance of analytical results for regulatory action, it is necessary to repeat the entire series of analyses through an official method.

The methods used in the laboratories vary considerably in time, complexity, and expense. The agency's goal in recent years has been to develop a reliable test for laboratory use, and then devise a simple, convenient form for screening, first in the laboratory, and later at the slaughter plant or docksite. Many of the standard methods of testing for compounds are laborious, expensive, and slow. Standard methods have the great advantage of being definitive, of providing state-of-the-art results for regulator use. However, quick, inexpensive, relatively sensitive screening tests can inform regulators and producers of specific problems, and can provide much valuable information within a short span of time.

For example, a technology of growing importance to FSIS is the immunological assay, particularly in the matter of chloramphenicol (CAM). Chloramphenicol is used effectively against a wide range of microorganisms, e.g., *Salmonella typhi,* other salmonellae, *Clostridia,* and *Listeria.* Because of definite evidence of serious bone marrow effects, the medical community reserves CAM use for those patients with infections where it has been shown to be clearly advantageous, and for infections that have failed to respond to other therapy.

Chloramphenicol is not approved for use in food animals in the USA. The only approved use in this country at this time is for medicating dogs and cats. The drug is readily available to the veterinary community and is illegally used on occasions to treat meat animals. Some nations exporting meat to the USA also legally allow CAM use in medicating food animals. The potential, therefore, exists for CAM residues in the meat supply. In response to FSIS/FDA concerns that CAM was being used illegally in bob veal sent to slaughter, FSIS conducted an exploratory survey at several slaughter plants. The CAM card test, an enzyme-linked immunosorbent assay (ELISA) specific for CAM at low parts per billion concentrations and rapidly performed on site, was used in plants to test slaughtered calves' urine for CAM residues. More than 300 calves were tested over a two-month period. Urine samples from several calves were CAM-positive by the card test which could not be confirmed in tissue by chemistry methodology at that time. To resolve this problem, FSIS chemists developed more sensitive gas chromatography/mass spectrometry analysis (GC/MS) methods for CAM in urine and muscle. The CAM calf project was conducted to verify that the new GC/MS methods would indeed confirm CAM card positive urine findings. Data derived from the CAM calf study confirm that now all positive inplant CAM card urine tests will be confirmed by the new GC/MS methods. The CAM card urine test was implemented at selected bob veal slaughter plants in 1990.

Federal meat inspection regulations require that prior to condemning edible meat for biological residues, that residue must be identified and quantified to confirm the violation. FDA regulations require that before prosecuting a producer for misuse of antibiotics, the samples must be analyzed and confirmed in violation by validated definitive methodology. The CAM card test is acceptable for screening samples for CAM and confirming CAM residue, but cannot be used for definitive quantitation, or estimation.

3.4 LABORATORY QUALITY ASSURANCE

This is a system providing assurance that a quality control system is performing adequately. These activities are performed through the review and evaluation of analytical data, work sheets, interlaboratory check sample programs, split sample programs, on-site laboratory audits (reviews), the setting of performance standards for both methods and analysts, and the determination of sample acceptability criteria and analyze stability in matrix. All of these activities must be properly documented. The provision of manuals and training is an integral quality assurance function (USDA, 1987a; USDA, 1988a; Garfield, 1984).

3.5 CRITERIA FOR COMPOUND EVALUATION AND SELECTION

It is not feasible to monitor for residues of all chemicals in meat and poultry. Further, this is not necessary to adequately protect public health. It is, however, important to assess the likelihood that animals exposed to chemicals may contain residues at levels of concern, and to conduct monitoring, where test methods are available, for those chemicals that are most likely to present the greatest potential risk. A hierarchical compound-assessment scheme is used for this purpose (USDA, 1988b, 1988c).

Each compound is evaluated on a number of factors so as to judge the potential for animal exposure and the significance for human health. These factors include:

- amount of actual or probable use
- conditions of use as related to residues at slaughter
- potential for misuse to result in harmful residues
- metabolic patterns of the chemical in animals, plants, and the environment, including the bioavailability and persistence of residues
- toxicity of the residue

Compounds are selected for monitoring and are included in a plan for the calendar year based on several factors, such as:

- compound ranking assigned
- whether a practical test method is available and is suitable for regulatory use
- whether the compound is measurable in a multiresidue method where many compounds, even though all may not be assigned a high ranking, can be tested at a relatively low cost
- monitoring or other experience that shows whether adulterating residues are present in meat and poultry

FSIS works from a list of about 400 compounds that includes certain environmental contaminants in addition to animal drugs and pesticides (USDA, 1988d). At present, FSIS has suitable regulatory methods of analysis for 162 of these compounds (USDA, 1988d).

Some compounds are routinely included in monitoring on a cyclical basis to confirm periodically that a potential residue problem does not exist. Cycling of compounds in monitoring allows the agency to include more compounds in the program than would otherwise be possible within its resources. Compounds rotated out of the program for a specific year are not disregarded; if the need arises, they can be added during that year. Over the last 10 years, virtually all the residues for which a suitable method was available have been monitored, except when a compound had an especially low ranking.

The process of compound evaluation and ranking is a dynamic one. Previous rankings may be affected by considering additional compounds, change in agricultural use, and new information on a compound's toxicity and its potential for leaving residues. The agency uses an advisory board of scientists from EPA, FDA, and USDA (FSIS and the Agricultural Marketing Service) to identify those points that may affect a compound or ranking, or indicate an urgent need for monitoring. In 1988, FSIS/Science announced the implementation of a prototype compound evaluation system (CES). The CES was designed to provide the agency with a more systematic approach to the categorization of compounds with respect to their likelihood of occurrence in meat and poultry, and their potential impact on public health.

Briefly, the CES addresses the risk of residues in meat and poultry as a a function of two major elements: *hazard* (adverse effects that may be produced by a given compound) and *exposure* (residue level; factors affecting level, such as use patterns, withdrawal times, etc., duration of or frequency of consumption of product containing residues of concern). The system is a two-value, hierarchical compound ranking scheme that classifies a given pesticide, animal drug, or contaminant in any one of 16 categories. Compounds of greatest concern carry a designation of A-1, (high health hazard potential; high likelihood of residue occurrence); those compounds of least concern are designated D-4 (negligible health hazard potential; low likelihood of residue occurrence). Care is taken to avoid the use of exact numerical rankings that might suggest a high degree of sophistication, which is possibly not justifiable because of data limitation, or the assumption inherent in the ranking process.

FSIS believes that the compound evaluation system is sufficiently flexible to permit rapid response to new information that may affect previous rankings, and to allow for the use of scientific or expert judgement. However, it must be emphasized that the CES was neither designed nor intended for use in the development of formal quantitative estimates of risk from meatborne residues. Rather it provides a rational basis for changes in compound emphasis with the National Residue Program and encourages development of new analytical methods for important compounds for which no methods exist. As such, the CES serves as a useful guide in the planning and allocation of FSIS program resources for those residues considered to represent the greatest potential effect on public health. The CES is updated as appropriate to provide the FSIS with a constant, informative, and sound approach to dealing with residues in meat and poultry.

3.6 STATISTICAL DESIGN CRITERIA

The identification of residue problem areas, i.e., determining which residues seriously involve which kinds of animals and therefore require special attention, is fundamental to achieving the goal of the National Residue

Program to reduce residue violations as much as possible. For this reason, the monitoring plan is based on a *residue/species pair* design concept. The procedures used to select compounds for monitoring have already been described. The species or production-class groups used in the pair-based examination are determined by common aspects of rearing, as these factors affect exposure and the likelihood of residues remaining at slaughter. For example, market hogs have an exposure-potential profile different from that of boars and sows.

For major species or production class groups whose food products are frequently consumed in substantial quantities by humans, sample numbers are selected to provide 95% probability that a residue problem will be detected if it occurs in 1% or more of the population. To achieve the stated level of assurance, a minimum sample size of 300 is required. The sampling plan can be modified to increase or decrease the desired assurance on the basis of previous experience with the residue (Table 3.1). In some cases we take more than 3,000 samples to evaluate a potential problem (USDA, 1988d; Parzen, 1960).

Although the monitoring program is not designed specifically to provide statistical estimates of the percentage of violations in large populations, such estimates are available as auxiliary information provided that a high degree of precision is not important.

The determination of the violation rate does not by itself accomplish the primary purpose of the agency to protect the consumer from products adulterated by residues. Violative results indicate a flaw in the control of residues at some point before FSIS involvement. Thus, upon receipt of a violative result, FSIS takes steps to identify the source. Follow-up actions normally require that additional animals from the farm be retained at slaughter pending tests to confirm that adequate remedial action has been taken.

TABLE 3.1. Numbers of Samples to Detect a Problem of Specific Size with Specific Probability.

Percent Violative Population	Probability of Detection		
	95%	99%	99.9%
1.0	299	459	688
0.5	598	919	1.379
0.1	2,995	4.603	6,905
0.05	5,990	9,209	13,813
0.01	29,956	46,050	69,075

3.7 EDUCATION AND RESIDUE AVOIDANCE

In addition to the sampling activities of the National Residue Program, with their attendant managerial and technological features, the agency has set up an educational and cooperative residue prevention enterprise called the Residue Avoidance Program (RAP). The residue avoidance concept is based on the notion that enforcement is more effective when combined with a cooperative educational program and communication with the involved groups.

Perhaps the most promising area of our endeavors is an initiative that entails government and industry sharing the responsibility and burden of assuring consumers that their food is safe and wholesome. This is a voluntary residue control agreement signed between FSIS and the federal slaughtering establishment that has official responsibility for the animal or poultry production. The agreement requires producers for the establishment to control all elements of production to prevent drug and chemical residues. FSIS monitors the agricultural practices, and verifies that the production controls are being applied as agreed. The producers utilize USDA-accredited laboratories, at their expense, to analyze feed and other products for contamination before using them on the farm. The producer also has a representative number of animals or birds sampled prior to presenting the entire flock or herd for slaughter. In the case of large food animals, urine or blood may be used in lieu of tissue samples. The establishments perform from 10–30 times the number of tests normally conducted by FSIS.

The first agreement (Memorandum of Understanding – MOU) was signed with a turkey establishment in 1976. RAP has renewed interest in the poultry industry, and stimulated new interest in the red meat industry. FSIS has not had many agreements with food animal producers, as the red-meat industry is less vertically integrated than the poultry industry. In order to expand participation, FSIS will soon propose regulations describing conditions under which integrated and nonintegrated companies alike may participate in a voluntary residue control (VRC) program.

MOU's are currently in force with 11 companies involving 29 slaughtering establishments. They represent a combined slaughter of 4 billion pounds per year. By commodity, the MOU's cover about 45% of young turkeys, about 19% of young chickens, and about 3% of feed cattle. The agency has received requests for participation from many others that could be considered when the VRC rule-making process is completed. The program has been very successful; with so much of the poultry production involved, violative residues in poultry have declined dramatically. Residue violations are now rarely found in an industry that once had repeated residue problems.

The three basic operating procedures of the program are monitoring, surveillance, and exploratory subprograms (Cordle, 1987, 1988; Engle, 1988; Houston, 1986; NRC, 1983, 1985; USDA, 1987b; 1988b).

3.8 ENFORCEMENT PROGRAMS

There are three phases to the regulatory programs that impact on the residue program effectiveness. *Monitoring* is conducted to obtain information on the frequency and levels of residues occurring nationwide over time. Area monitoring may be conducted where a localized potential problem appears. Monitoring information is obtained through statistically based selection of random samples from healthy-appearing animals under inspection. The samples are then analyzed for residues based on an annual plan (USDA, 1988d). The data are used to evaluate residue trends and to identify problems within the industry where educational or regulatory action may be needed for correction. Thus, monitoring not only gathers information, but also deters practices that lead to violative residues. *Surveillance* is designed to investigate and control the movement of potentially adulterated products. The sampling is biased and is directed at particular carcasses, producers, buyers or products, etc., in response to information from monitoring or other information sources, or from observations during ante- or post-mortem inspection indicating that adulterating levels of residue may be present.

The *exploratory* phase is conducted for a variety of reasons, other than monitoring or surveillance, such as increasing information about specific compounds and their use, or establishing data on the transfer of naturally occurring toxins from feed to livestock.

In addition, FSIS has implemented a nationwide database, the residue violation information system (RVIS), to review, sort, cross-reference, and manage all residue data obtained by FSIS, FDA, or other agencies from residue violation cases. This includes names and addresses of sellers and producers, dealers, and the results of investigations.

USDA has also implemented a swine identification program to trace hogs back to their farm of origin. A number of industry and consumer groups have urged the department to broaden the program to include other species, not only to control residues but to control animal disease as well.

3.9 CONCLUSION

In the years that lie immediately ahead, maximum use will be made of residue avoidance and cooperative residue programs directed toward generic prevention of problems through appropriate controls at all critical stages of production. The agency will allocate its resources to particular compound/species pairs, identifying through improved methods of risk management where and how to target its efforts. Surveillance efforts will be directed toward the small percentage of industry that is marginal in performance.

FSIS will continue to improve its testing capability to be able to detect more residues more efficiently. Major interests will be in rapid, cost-effective screening methods for use on the farm, in the plant, or at dockside; multi-residue laboratory procedures for major compound classes; and improved sensitivity of methods to provide better information on residue frequency distribution for risk management purposes.

In the effort to standardize international regulatory controls, there will be a continuing need for exchanges of scientific information and acceptance of shared data among trading nations. But ultimately—as illustrated by the differences between the U.S. and the European Economic Community on the use of anabolic steroids—more must be done to advance the understanding of the consumer.

By 1995, we hope to see a residue situation that will permit FSIS to devote its major residue efforts to education and communication with industry and consumers. FSIS would then be acting primarily as a quality-assurance program for the U.S. meat and poultry supply. The fundamental situation we envision would offer industry the benefits of objective verification of good quality meat or poultry products and freedom from economic loss; the consumer would enjoy the assurance of a safe and wholesome food supply.

3.10 REFERENCES

AOAC. 1984. *Official Methods of Analysis*. 14th ed. Arlington, VA: Association of Official Analytical Chemists.

Cordle, M. 1987. "Residues and What FSIS Is Doing to Regulate Them in Meat and Poultry Products," presented at the annual meeting of the Southern Association of Agricultural Scientists, Nashville, TN, February 3.

Cordle, M. 1988. "USDA Regulation of Residues in Meat and Poultry Products," *J. Anim. Sci.*, 66(2):413.

Engel, R. E. 1988. "Quality Assurance Through Residue Screening," 192(2):268.

Garfield, F. 1984. *Quality Assurance Principles for Analytical Laboratories,* Arlington, VA: Association of Official Analytical Chemists.

Houston, D. L. 1986. "Developments in the Residue Program at USDA-FSIS," *Food and Drug Cosmetic Law Jour.*, 41:438.

National Research Council. 1983. *Risk Assessment in the Federal Government: Managing the Process Committee on the Institutional Means for Assessment Risks to the*

Public Health Commission on Life Sciences. Washington, D.C.: National Academy Press.

National Research Council. 1985. *Meat and Poultry Inspection—The Scientific Basis of the Nation's Program.* Washington, D.C.: National Academy Press

Parzen, E. 1960. *Modern Probability Theory and Its Application.* 1st ed. NY, NY: John Wiley & Sons.

U.S. Code of Federal Regulations. 1987. Title 21, Parts 500-599, p. 1.

U.S. Code of Federal Regulations. 1985. Title 40, Parts 162, p. 73.

USDA. 1987a. *Chemistry Laboratory Guidebook (Revised).* Washington, D.C.: FSIS, Science Program.

USDA. 1987b. *Microbiology Laboratory Guidebook.* Washington, D.C.: FSIS, Science Program.

USDA. 1988a. *The National Residue Program.* Washington, D.C.: FSIS, Facts-18.

USDA. 1988b. *Compound Evaluation System.* Washington, D.C.: FSIS, Science Program.

USDA. 1988c. "Compound Evaluation System; Availability, " *Federal Register*, 53:751.

USDA. 1988d. *Compound Evaluation and Analytical Capability—National Residue Program Plan.* Washington, D.C.: FSIS, Science Program.

CHAPTER 4

Methods of Detection

RICHARD ELLIS, Ph.D[1]

4.1 INTRODUCTION

ONE objective of many public health agencies is to ensure safe and wholesome products of animal origin for consumers. Assurance of this priority can no longer only be assumed by the careful ante- and post-mortem examination of animals, the facilities in which food animals are slaughtered, or monitoring the formulation of processed products.

During the last half-century there have been dramatic increases in the use of drugs and pesticides in the production, storage and transportation of food. The meat and poultry industry is no exception. This growing use of drugs and pesticides has inevitably led to concerns of adulteration of food products caused by persistent residues. Awareness of adulteration in food-producing animals and poultry has expanded as use of these chemicals has proliferated throughout the world. Today, for a number of sound reasons, a large percentage of livestock and poultry intended for food production may be exposed to many chemicals.

Residues of these chemicals cannot be detected by traditional means of inspection. Scientific breakthroughs have led to many new, selective, and sophisticated products. The proliferation of products has been accompanied by the development of new techniques that have made it possible to detect very low levels of veterinary drug and pesticide residues in food.

The full consequences of pesticide and drug residues become clearer as new detection methods are developed and mature. Public health scientists

---

[1]Director, Chemistry Division, Food Safety and Inspection Service, USDA, Washington, D.C., USA.

are finding that some of these chemicals have become ubiquitous in our environment, and that they persist for longer periods than originally expected. Today, the evolution of innovative analytical procedures and instrumentation allow an ever broadening spectrum of residues to be detected including chlorinated hydrocarbon, carbamate and organophosphorus pesticides, antibiotics and antimicrobials, sulfanamides, anthelmintics, coccidiostats, hormonally active drugs, other therapeutic and prophylactic drugs and industrial chemicals. Thus, methods of detection now play a substantial role in this public health arena.

4.2 NEED FOR REGULATORY ANALYTICAL METHODS

Responses by national governing bodies to the constituencies they serve have resulted in a growing list of regulatory requirements and needs. In responding to the growing armada of chemicals for animal and poultry products intended for consumption, regulatory agencies must provide measures to ensure that these potent chemicals will not have adverse health effects on either animals or consumers. It is enlightening that science has had a positive influence on many of these public health regulatory developments.

Public laws and regulations have evolved that require more scientific information be made available to evaluate food safety issues such as carcinogenicity, mutagenicity, immunosuppression, allergenicity, and residue persistence. Documented histories of some selected drugs and pesticides such as DDT, diethylstilbestrol (DES) and chloramphenicol argue for such information. The growing issue for certain antibiotic and antimicrobial agents leading to drug-resistant bacterial strains may add to this list of drugs and pesticides.

The need for data to address scientific questions requires that fundamental information on the residues in animals be provided, in part, by analytical methods. Clearly, analytical methods provide only part of the data needed. Another major contributor for these in-depth evaluations is basic toxicology. The integration of these disciplines provides much of the qualitative and quantitative data needed to fully evaluate, and ensure the safe use of, drugs and pesticides.

In most countries today, a demonstrated practical analytical method to detect residues of these drugs and pesticides is required as a part of the registration process. Outside the U.S. and Canada this requirement is, for example, laid down in a European Economic Community (EEC) directive. The Federal Republic of Germany has implemented a corresponding paragraph in the German Drug Act as amended in February 1983. This requirement to provide the analytical method for residue analysis now usually falls upon the drug or pesticide sponsor. Thus, the need for adequate analytical methods transcends the national health agencies to the animal health industry.

The structural identity of a biologically transformed drug or pesticide, as well as its quantity, are essential parts to understanding mechanisms of action and metabolic depletion in animal tissues and biological fluids, and form a basis for conducting appropriate toxicological studies. Data from such studies help the appropriate public health agencies establish the necessary safe use and allowable residues of these drugs and pesticides for consumers, food-producing animals and poultry. With a notable number of older drugs and pesticides, many of these current requirements did not exist at the time of their registration, and so it falls upon the public health agencies to develop the necessary analytical methods.

4.3 ATTRIBUTES OF REGULATORY ANALYTICAL METHODS

National regulatory agencies and programs want effective and practical analytical methods that can routinely detect, reliably quantify, and unambiguously identify all residues of pesticides and drugs that may be present in meat and poultry at the appropriate level of interest. Methods with these attributes are not available for many of the compounds of interest, in part because of the extensive number of potential residues that may find their way into the food chain.

A method suitable for regulatory control purposes has to be reliable. To ensure analytical method reliability, performance characteristics of a method must be determined by multilaboratory evaluation. In most cases, minimum standards should be designed to fit the needs of specific program requirements. Regardless of specific analytical requirements, all methods are characterized by a set of attributes that determine their applicability. By consensus with many public health standard–setting organizations or agencies, the principal attributes of analytical methods are specificity, precision, systematic error (an analytical method bias), and sensitivity. Other attributes also contribute to identifying acceptable analytical methods.

Specificity is the ability of a method to respond only to the substance being measured. The proposed method must provide for the identification of the compound being measured. This characteristic is often a function of the measuring principle used and the function of the compound under study. A key consideration of specificity is that it must be able to differentiate a compound quantitatively from homologues, analogues or metabolic products of the residue of interest under the experimental conditions employed.

Precision is a measure of the variability of repetitive measurements with the method when applied to separate portions of a homogeneous sample. Factors affecting precision are the variability associated with analytical results from different laboratories—defined as reproducibility—and the variability from repeated analyses within a laboratory—defined as repeatability. Precision is usually expressed as a standard deviation, and in some standard-setting organizations, this term is used to describe other method characteristics such as limit of detection, limit of decision (EEC Commission, 1987) or limit of reliable measurement. Another useful term is the relative standard deviation, or coefficient of variation, because it is relatively constant over a considerable concentration range (an order of magnitude, for example), ideally covering the level of interest.

The variability achieved in the developing laboratory after considerable experience with the method is usually less than that achieved by less-experienced laboratories who may later use the method. For this reason, the final version of the method should be optimized by procedures described by Youden and Steiner (1975). If a method cannot achieve a suitable level of repeatability in the developing laboratory, it cannot be expected to do any better in other laboratories.

Systematic error is an analytical method bias. It is the difference of the measured value from the true, assigned or accepted value (target value). It is generally expressed as the percent recovery of analyte added to a blank tissue matrix. In many cases of residue analysis, the analyte added to a sample may not behave in the same manner as the same analyte biologically incurred. At relatively high concentrations, and particularly with methods involving a number of steps (extraction, isolation, purification and concentration), percent recoveries may be lower. Regardless of what average recoveries are observed, recovery with low variability is a desirable feature.

Accuracy is usually synonymous with systematic error. Accuracy of the mean, for example, refers to the closeness of agreement between the true value, and the mean result that would be obtained by applying the experimental procedure a very large number of times to individual samples of homogeneous material. The accuracy requirements of different types of methods will vary with the use being made of the results. Generally, however, accuracy must be at least equal to or greater at or below the level of interest than it is above. This does not imply that accuracy is unimportant above the level of interest, e.g., a tolerance or maximum residue level.

The sensitivity of a method is a measure of the ability to discriminate between small differences in analyte concentration. A common practice is to define sensitivity as the slope of the calibration curve with known standards at the level of interest. An ideal situation would be afforded by a linear curve.

Beyond these principle method characteristics are a number of collateral criteria suitable for analytical methods for regulatory control programs.

Methods should be rugged or robust, cost effective, relatively uncomplicated, portable and capable of handling a set of samples simultaneously in a reasonable period of time. Ruggedness of a method refers to its performance being relatively unaffected by small deviations from the established values in the use of reagents, quantities of reagents used, or temperature/time factors for extractions or reactions. This does not provide latitude for carelessness or haphazard techniques. Cost-effectiveness refers to use of relatively common reagents, and avoiding instrumentation or equipment not customarily available in a laboratory devoted to trace environmental analyses. Relatively uncomplicated methods use simple, straightforward mechanical or operational procedures throughout. Portable methods can be moved from one location to another without loss of established performance characteristics. The capability for simultaneous analyses aids in method efficiency by allowing sets or batches of samples to be analyzed at the same time. This attribute reduces time requirements for sample analysis to eight or more analyses in a normal working day. This is particularly important when large numbers of samples must be analyzed.

These characteristics of method performance should be determined at or below a certain concentration of an analyte, or a level of interest established by the responsible regulatory body within an individual country, group of countries, or standardizing body. Clearly, the more uniformity there is in defining these values, the more universal appeal a method may have. In fulfilling the above criteria, a positive result affirms (or proves) the presence of the analyte in a sample, according to the method. A sample result should be considered quantifiable if the criteria specified for the relevant method are not fulfilled, or the result does not indicate the presence of the analyte in the sample above a limit of decision or limit of reliable measurement.

The importance of establishing the attributes and performance criteria cannot be over-emphasized. Analytical results from methods with performance standards provide the necessary information to allow regulatory control officials to develop and manage programs responsive to their public health responsibilities. Performance criteria for analytical methods also provide a basis for good management decisions in future planning, evaluation and product disposition. For the animal health industry, it provides a guideline for knowing what performance characteristics an analytical method must achieve in developing analytical procedures for registering new animal drugs and pesticides with regulatory authorities.

4.4 CATEGORIES OF ANALYTICAL METHODS FOR REGULATORY PROGRAMS

A universal analysis scheme that can simultaneously quantify all compounds or classes of compounds of interest in animal tissue or fluid, with acceptable accuracy, and which would correctly identify the analyte or analytes, would provide a desirable and unified method approach. A number of samples could be analyzed in a reasonable time frame, and consume no more time than required for screening procedures for the presence of an analyte. This scenario reduces the need for single analyte procedures and the need for performing the numerous extraction-based methods now used to determine residues of individual analytes. It would not minimize the need for the very important multiresidue microbial bioassays. Rather, these two approaches would complement each other. Nonetheless, at the present time there are very few analytical procedures available to regulatory agencies that simultaneously quantitate and confirm the identity of pesticide and veterinary drug residues in animal tissues or fluids. Until more universal methods are available, regulatory programs must employ methods with individual attributes of presumptive presence, quantitation and positive identification.

Regulatory control and standard-setting bodies have used several different terminologies to describe analytical characteristics of methods. Terms such as *confirmatory, reference, defensible, routine, quantitative, semiquantitative, screening, rapid* and *presumptive* methods are well known. An alternative to the potential difficulty of categorizing methods, and the stigma associated with these descriptive terms, is to define the methods independent of intended purpose according to their attributes, or qualities of method performance. Attributes and qualities of three levels of analytical methods are relevant to support regulatory control programs.

Level I methods incorporate the ability to quantify the amount of a specific analyte or class of analytes and positively identify the presence of an analyte in a single analytical procedure. These are assays with the highest level of credibility, and are unequivocal at the level of interest. They may be single procedures that determine both the concentration and identity of the analyte, or combinations of methods for determining and confirming a residue for definitive identification. Frequently, chemical confirmatory methods employ a chromatographic technique combined with mass spectrometry (Burlingame, 1988). Detection and identification by mass spectrometric methods often employ selected ion monitoring and the ratios of selected ions (Sphon, 1978). Although Level I methods rely on instrumental procedures, observation of a lesion or other morphological change which is pathognomonic for a specific disease, or exposure to a class of veterinary drugs or pesticides, could be a Level I method. The potential concern with such bioassays is their sensitivity and precision. Level I methods may also be limited to analytes with appropriate physical and chemical properties. For example, at the present time there are few antibiotic drugs that have mass spectrometric procedures applicable for use in a regulatory control program.

Level II methods are those which are not unequivocal, but are used to determine the concentration of an analyte at the level of interest, and to provide some structural information. For example, these methods may employ structure, functional group or immunochemical properties as the basis of the analytical scheme. These methods are often reliable enough to be used as reference methods. Level II methods commonly separate the determinative assay from the positive identification procedures, and may also be used to corroborate the presence of a compound or class of compounds. Thus, two Level II methods may provide attributes suitable for a Level I method. The determinative part is equivalent to the quantitative portion of the Level I methods, and may provide some identification parameters. For example, certain types of compounds may be excluded or included in the extraction procedure, or specific elemental detectors may provide inferential information. The majority of analytical methods presently available and used by regulatory control agencies are Level II methods.

Level III methods are those that generate imperfect, though useful, information. These testing procedures detect the presence or absence of a compound or class of compounds at some designated level of interest, and often are based on noninstrumental techniques for analytical determination. Results on a given sample are not as reliable as Level I or II methods without corroborating information. Level III methods may provide reasonably good quantitative information, but poor compound or class specificity or identity, or may provide strong or unequivocal identification with very little quantitative information. Level III methods are not poorly described or sloppy methods, rather, they have defined operating characteristics of reliable performance. Many of the microbiological assay procedures and immunoassay-based methods may fall into this category. They are commonly used because of their greater sample capacity, portability, convenience and potential suitability to nonlaboratory environments. The hallmark of Level III methods is action based on individual positive results which require verification, using Level I or II methods as required by the uncertainty of an individual result. However, epidemiological information may provide substantive data, reducing the uncertainty of individual results. To a regulatory control program, these methods may offer substantial advantages, including analytical speed, sample efficiency through batch analysis, portability to nonlaboratory environments, sensitivity, and the ability to detect classes of compounds. Even though Level III methods may not detect specific compounds at regulatory limits on all samples, they are able to test larger numbers of animals and poultry carcasses with a limited level of resources.

The reliability of Level III methods should be measured in part by their performance characteristics, as well as their ability to handle relatively large numbers of samples within a given time frame. Two key characteristics requiring definition include the percent false positives and percent false negatives when measured against a validated quantitative assay in a statistically-designed protocol. The percent false negatives must be quite low at the levels of interest, while slightly more flexibility may be acceptable for false positives. A minimum residue detection limit can be described based on these two parameters.

Reliability for Level I methods must be restrictive. False positives and false negatives should be at or near zero. For Level II methods, false positives and false negatives should approach Level I methods, particularly for those methods that are used in regulatory control programs having evidence of residue violations. Level III methods can produce a low percentage of false positives, because a Level I or II method will commonly be used to verify and quantitate results. Percentage of false negatives must be very low for regulatory programs.

4.5 METHOD DEVELOPMENT AND VALIDATION PROCEDURES FOR REGULATORY METHODS

Besides the development of the analytical method itself and optimizing its performance, the most important factor in defining a method's performance characteristics is the multilaboratory validation study.

In developing a regulatory method, originators need to collect data from three types of samples:

(1) Control tissue from nontreated animals
(2) Fortified tissue samples containing known concentrations of the analyte added to the control tissue
(3) Dosed or incurred tissue from animals that have been treated with the drug or pesticide

These tissue samples enable the originator to define the baseline response that the method can satisfactorily recover, to identify known amounts of the analyte added to the control tissue, and to demonstrate that the method can satisfactorily recover the biologically-incurred residue.

Residue methods should be designed with as much simplicity as possible to limit the variety, size and type of glassware and equipment needed, to minimize the potential for analytical error, and to reduce costs. Reagents and standards must be readily available. Specific instrumentation should be based on performance characteristics rather than a particular manufacturer. A desirable feature is analytical instrument flexibility to control laboratory instrument costs.

Subjecting methods to different residue-testing environments may place some additional requirements on methods. Warmer environments require reagents to be more thermally stable, solvents to be less volatile, and tissue sample considerations to be more tolerant. Cooler environments may require reagents and solvents with physical properties such as lower freezing points and higher solvating properties to ensure effective extraction of an analyte. Environmental temperatures may also influence the time required to perform an analysis, as well as such phenomena as reaction rates for derivatization and color development. These considerations may strain efforts to standardize methods in broadly differing environments, because of the need to adapt methods to different environmental factors.

As requests for analytical residue determinations are pushed to lower limits of detectability, analytical sensitivity becomes critical. Analytical reproducibility becomes more difficult to attain, and coefficients of variation increase. Performing replicate analyses improves the reliability, but for regulatory programs, this reduces the number of samples that can be analyzed, and increases the analytical cost for each reported analytical result. In addition, confidence of scientific credibility and disposition of product is lessened, and may be seriously challenged by producers, meat processing establishments, consumers, consumer action groups, or other regulatory control agencies as analytical determinations are pushed to lower limits. These factors reinforce the need for method validation procedures.

An analytical method developed and used in only one laboratory is of limited use, because the credibility of reported values may arise even though strong quality control procedures may have been employed. As a minimum, three laboratories expected to use such methods should be able to successfully conduct the analytical procedure and obtain statistically acceptable agreement on the same samples split among the testing laboratories. Methods with high reliability for residue testing should be able to successfully undergo a collaborative study involving at least eight different laboratories (Horwitz, 1988).

Samples included in either a validation or collaborative study should contain the residue at concentrations bracketing the acceptable residue limit and include blank tissue as well. All study samples should be analyzed over a limited number of days, and at least in duplicate, to improve statistical evaluation of method performance. These are only minimal requirements. Duplicate analysis and tissues would yield limited quality estimates of repeatability and reproducibility.

The principles of conducting either a validation or collaborative study of a method are the same. Considerations include informing participants of the study and involving them in the study design. Laboratories should be chosen with as wide a distribution as possible, and should include regulatory control agency, international, industry and academia laboratories. Expected performance standards for both the method being studied and for the performing analysts are advantageous, and should be sent before the study begins. Sample shipment techniques should be prearranged and should include practice samples. This is particularly important when studies are conducted internationally. Prudent and thorough planning are essential for these studies because of their expense.

4.6 QUALITY ASSURANCE AND QUALITY CONTROL

Regulatory control agencies responsible for monitoring residues in food are routinely reminded that they conduct programs in a global fishbowl. Any analytical discrepancy requires the inevitable defense of an analytical program. A principle objective, therefore, becomes one of assuring that there is a well-planned and executed quality assurance program.

Quality assurance begins with the method development process. Activities include experimental optimization of each procedural step or manipulation to determine the critical control steps—those having a substantial impact on method performance. Other activities include identifying where an analytical method may be stopped without adversely affecting the result, determining the ruggedness or process variability that may be employed in any particular method step without reducing the method's performance, and determining the tissue requirements necessary to ensure reliable analytical performance by conducting short-term residue stability tests in tissue matrices. Instrument parameters, and procedures to test instrument performance, need to be documented as part of the method. Mass transfers in the procedure should be minimized. Finally, the method must be written in thorough, concise, unambiguous language. A list of the reagents and supplies, as well as their commercial sources, must be included. Instrument performance parameters and conditions have to be documented. These factors will facilitate method transfer into a regulatory program.

Unfortunately, the extent of commitment to a quality assurance program will vary from laboratory to laboratory. It will depend on the complexity of the operation, the purpose for which the laboratory exists and the willingness of management to commit resources. A high degree of attention to quality is essential in a regulatory control agency and its enforcement activities. Smaller laboratories can operate with smaller programs, but supervisory personnel must know what is necesary to achieve high quality laboratory results.

An integral part of quality assurance is a quality assurance plan (QAP) for each analytical method performed in the laboratory. Examples of quality assurance plans are available from some regulatory control agencies (USDA, 1982). The objectives of these plans are: to specify quality-related activities, such as training; to specify requirements to ensure reproducible and valid data and results; to help analysts identify activities that can affect the validity of results before analytical work begins, and to insure that critical operations are identified and verified.

These plans provide a systematic scheme to document the quality of a determination. A QAP should consist of five parts:

(1) The performance standards of the method (Analyst performance standards are established separately and may vary among different laboratories.)
(2) The critical control points of the method
(3) Analyst training
(4) Interlaboratory and intralaboratory check samples
(5) Uniform, well-defined materials used as analytical standards

Analyst training, or determination of readiness to perform analyses, should be described in a familiarization protocol consisting of: training on pure standards at levels of interest; analyst fortified samples for recovery with the test matrix, concentrations and numbers of replicates and blind check samples using concentrations of the analyte(s) of interest, which are unknown to the analyst.

Many regulatory control agencies require an analytical method as part of the veterinary drug registration process. In presenting a proposed regulatory method, the originator must provide many of the parameters noted previously, such as: a complete written description of the method performance; typical standard curves prepared from the analyte of known purity and fortified control tissue; data on specificity, precision and systematic error attributes; relevant work sheets, statistical analysis of the data; suitable spectra and/or chromatograms from control, fortified and incurred tissue. This information facilitates the method approval process.

The focus on quality results should be emphasized. It is, in the long-term, less expensive to do it correctly the first time. This ensures credibility to a regulatory control program and esprit de corps among analysts.

4.7 EXAMPLES OF REGULATORY CONTROL METHODS

Concerns about veterinary drug residues have grown in scope over the last two decades. Today it is recognized as an issue that transcends national boundaries. The complexity of factors involved led the Codex Alimentarius Commission (CAC) to convene a group of experts to provide guidance on the problem. The consultation report recommended establishing a Codex committee to address this international problem (FAO, 1985). At its first meeting, the Codex Committee on Residues of Veterinary Drugs in

Foods identified a list of drugs needing highest consideration for establishing maximum residue levels (Codex 1987, 1988, 1989). Among the drugs considered were the synthetic and natural anabolics, chloramphenicol and sulfonamides including sulfamethazine. Descriptions of these three residue classes provide examples of applying the principles described earlier in this chapter.

The USDA's Food Safety and Inspection Service identified sulfamethazine residues in swine as a prevalent and continuing residue problem. In response, an effective Level II method based on thin layer chromatography with fluorescence detection and quantitation was developed. This procedure was successfully subjected to a collaborative study (Thomas, 1983) and has been used in their regulatory control program. With the continued presence of sulfonamide residues, a new method has been developed based on the quantitative sulfonamide test procedure. This Level III method employs the technology of the qualitative Level II method referenced above.

The technical approach for the method, which is now referred to as the sulfa on site (SOS) test, began with three correlation studies (Randecker, 1987). The primary focus of these studies was on correlation data for sulfamethazine. However, the method also detects other sulfonamides that are approved for use in swine feed. The data generated from these studies were used to establish ratios of drug residue concentrations in tissue to urine and tissue to serum. The ratio of drug residue concentrations of sulfamethazine for urine versus liver was estimated to be 4 to 1, and for urine versus muscle, 12 to 1. This allowed estimation of the concentration of sulfamethazine in the liver or carcass based on the concentration in these biological fluids. Data on tissue to fluid relationships are summarized in Table 4.1.

The simplified sulfamethazine method eliminates the extraction and cleanup procedures employed in the Level II method. Urine may be applied directly to a silica gel thin-layer chromatography plate with a thicker, preabsorbent layer. Two solvent systems are used to provide separation so

TABLE 4.1. Tissue:Fluid Ratios for Sulfamethazine in Swine.

	Muscle		Liver		Kidney	
	Ratio	CV*	Ratio	CV	Ratio	CV
Serum	0.24	55	0.90	79	0.53	21
Urine	0.08	122	0.27	130	0.16	87

*Coefficient of variation.
Source: Randecker, 1987.

that semi-quantitative estimates of sulfamethazine residues may be made by comparing the fluorescence of the unknown urine sample with 0.4 mg/kg and 1.3 mg/kg sulfamethazine standards. The test can be performed at low cost in less than 40 min. Color intensity judged to be less than the low standard is indicative of an animal carcass with sulfamethazine residues less than 0.1 mg/kg in all tissues. Color intensity judged greater than the low concentration standard, but less than the high concentration standard, is indicative of organ tissues greater than 0.1 mg/kg, and carcass residues less than 0.1 mg/kg. Urine concentration judged to be greater than the high standard is indicative of a carcass residue concentration greater than the 0.1 mg/kg residue tolerance.

Field trials have been conducted to evaluate and refine the SOS test procedure. Samples of urine were tested on-site at swine abattoirs in Ohio, Iowa, Georgia, and North Carolina. The urine, liver, and muscle tissues from the corresponding animals were submitted for laboratory testing. The data produced by these trials established the ruggedness and reliability of the SOS test.

In addition to these studies, a collaborative study was conducted on the SOS procedure using the self-instruction guidebook. The data from more than 1,500 analyses by 10 collaborators was collected for statistical review. Preliminary reports indicated that results of the SOS collaborative study corroborate well with previous data compiled from the controlled studies and the field trials.

A recently developed procedure by Henion and coworkers (Covey, 1987) provides a novel approach to residue analysis for diethylstilbestrol (DES) and zeranol, with performance characteristics and attributes that enable the procedure to be used either as a Level III method for rapid screening for DES and zeranol or, by adding an internal reference standard into the procedure, a Level II and Level I method is attainable. This flexibility is unique and very suitable for a regulatory program. The method has demonstrated the versatility to analyze kidney, liver, and muscle at less that 1 g/kg DES and less than 2 g/kg zeranol, with a coefficient of variation of 25% or less on unknown, incurred residue tissue samples (Henion, 1988). Recent developments suggest extension to trenbolone and melengestrol acetate.

The procedure starts with an enzyme hydrolysis followed by a three-phase solvent extraction process to remove the majority of the triglycerides and highly polar materials. An ion exchange cartridge is used to extract the DES and zeranol. Several solvent washes are performed prior to the final elution. The analytes in the purified extract are converted to trimethylsilyl derivatives and analyzed using gas-liquid chromatography for screening. If an internal standard is added to the test sample, quantitation and confirmation may be done when a mass selective detector (a simplified mass spectrometer) is coupled with the gas-liquid chromatography system. The manual extraction and purification steps are amenable to automation using a centrifugal column elution device, which significantly increases sample throughput. Successful application of this technology to other growth-promoting drugs provides a rapid, versatile multiresidue procedure for regulatory control programs, approaching the desired universal analytical scheme discussed earlier.

A promising, recently introduced approach to residue analysis involves application of immunochemical-based assays. Such procedures show clear promise of providing numerous Level III and Level II methods. When combined with other analytical schemes or criteria, Level I methods may be possible with this technique.

One such procedure for chloramphenicol was described by Mumma and Hunter (1988). The assay can be applied to the analysis of meat, milk, eggs, and various body fluids. The procedure is a saturation-based assay using the principle of mass action. The analyte and a radio-labelled analogue of chloramphenicol saturate a limiting amount of specific antibody binding sites. The fraction of bound, labelled analogue varies with the concentration of the analyte in a sample.

The antibody used in the assay was produced in sheep using antigens of chloramphenicol bound to a suitable carrier protein using the side-chain of the molecule. The only significant cross reactivity observed with this antibody was for side-chain metabolites of the parent compound. These metabolites, however, seem to play a minor role for residues in the edible tissues of the food animal species tested. This may be substantiated with the finding that results obtained with this radio-immunoassay (RIA) could be confirmed quantitatively by physico-chemical methods exhibiting selectivity for the parent compound. The extracting and purification procedures are short, and include a minimum number of steps. Plasma, bile, and saliva can be analyzed without any sample cleanup. The fat has to be removed from milk by extraction prior to analysis. The cleanup for meat samples includes extraction with an organic solvent, removal of lipids, partitioning into another organic solvent, evaporation and dissolution of the dry residue in an appropriate phosphate buffer. The method could be considered a Level II method and it fulfills most of the requirements of a Level I method except for the direct information on the structure of the analyte. This procedure has a sample throughput that is more typical of Level III methods.

The chloramphenicol RIA is used in the Federal Republic of Germany and it is being applied in other European laboratories (Arnold Somagi, 1982). It was studied collaboratively in the member states of the European Economic Community.

4.8 TRENDS AND EMERGING TECHNOLOGIES

There is strong temptation to separate trends in the analytical detection portion of residue control programs from emerging technologies. However, in a number of cases, implementing new regulatory control programs relies on the development, refinement and introduction of these new technologies. A case in point is the notable trend to rapid testing systems employing immunochemistry-based card assays now commercially available. For this reason, trends and emerging technologies are considered jointly.

New technologies generally fall into two broad categories. The first is biotechnology-based, while the second is related to traditional analytical chemistry and its associated disciplines such as physical chemistry. The first category is largely the outgrowth of genetic engineering and immunochemistry developments, while the second is the result of advances in electronics and material sciences.

Application of these technologies to animal drug and pesticide residues has both short-term and long-term aspects. In the short-term, a primary emphasis is to more fully utilize these emerging technologies, such as immunochemical procedures for animal drug residues. These procedures have been developed, and are well-established, in other areas such as clinical chemistry. Immunochemical and biochemical techniques used in the research arena of animal drug development may assist in transferring this technology into usable regulatory methods. In the arena of instrumentation and materials science, fuller utilization of developments in high performance liquid chromatography, such as supercritical fluid applications, robotic or automated analytical procedures, and integrated mass spectrometry procedures are most notable.

In the longer term, the major effort will be to identify and evaluate current research for potential applications to animal residue methods. Examples with potential include miniaturization of chemical detectors for on-line analysis, or interfacing these detectors with biochemical or immunosensors. Another major area to focus on is technologies that address extraction and purification procedures to eliminate potential interfering substances from tissue matrices, and feeds allowing detection of the small quantities of analytes by chemistry-based, rapid testing procedures. Clean extracts are also important for microbiological-based assays, but are generally not as serious a limitation to method development as they are for chemical-based assays. Current research has expanded to include sample preparation to facilitate applications to the newer, rapid test systems.

The development and application of immunoassay systems has been a notable accomplishment in recent years. Current techniques include RIA, enzyme-labelled immunosorbent assays (ELISA) and others (Mumma and Hunter, 1988). These techniques allow detection and, usually, quantitation for a broad range of molecular weight compounds including proteins, drugs, and pesticides at g/kg to ng/kg levels. They do not, at this time, necessarily provide high specificity, but it is possible to refine this property. The individual procedures do have some restrictions on instrumentation, yet are potentially capable of being Level II methods. Their advantages include their sensitivity, high sample capacity, rapid sample throughput and potential for automation. These qualities make the methodology attractive to new and improved regulatory control programs. Each individual approach has its individual disadvantages, such as a specific equipment needs and suitable antibody availability. However, the advantages outweigh the disadvantages. The approaches offer the advantage for developing testing systems usable in non-laboratory environments—another trend in regulatory control programs.

The instrument technique advances in liquid chromatography are most notable in applications to drug and pesticide residue testing. Improvements in liquid chromatography columns, fluorescence, electrochemical and diode array detectors, electronic controllers, data monitoring systems and the availability of solvent systems are broadly recognized. Advances in instrument design and incorporation of microcomputers provide valuable information for more specific structural and chemical information about analytes. This technology also provides new applications of infrared analysis to residues of drugs and pesticides in animal tissues. If reliability and sensitivity can be demonstrated, these instrument improvements provide a source for more and better Level II methods. The versatility of mass spectrometry as a technique for Level I, II, and III methods has been enhanced by the successful coupling of mass spectrometers to chromatographic systems as well as supercritical fluid extraction. Although these systems are relatively expensive, they do provide broad potential use for Level I and II methods that are cost-effective for regulatory control. This instrumental approach requires highly skilled and trained operators, but the definitiveness of the technique may be invaluable for regulatory programs. This potential versatility cannot be overlooked.

The need exists to improve and unify extraction and purification schemes for drug, pesticide and other environmental residues in food. One emerging technology with high potential for enhancing public health protection by regulatory programs is application of supercritical fluids, particularly when coupled with liquid chromatography technology (Charpentier and Sevenants, 1988). Supercritical fluid extraction using either pressure or temperature programming or suitable chemical modifiers has the effect of changing the solvating power and properties of the supercritical fluid. This versatile feature gives the technique the potential of being able to easily and selectively separate a wide variety of analytes not only from each other, but also from the tissue matrix in which they are found. Examples of applications such as these are being noted with increasing frequency. The application of this technology will be Level I and Level II methods as laboratory test procedures because of the special high pressure equipment needed.

The other notable trend in analytical chemistry technology is laboratory automation, ranging from instrumentation with integrated microcomputers to robotic systems capable of performing repetitive analytical procedures 24 hr/day. This approach continues to grow in scope and application as noted by the commercialization of instrumentation and international conferences. A recent development for time-saving separation and isolation of analytes from tissue extracts provides programmable, automated solvent dispensing capability. The centrifuge-based instrument mentioned previously allows use of smaller sample sizes and solvent volumes, while permitting analysts to purify and isolate analytes in batches of up to 12 samples/hr. This laboratory-based system is expected to have many potential applications to regulatory programs, and to lead to improved laboratory effectiveness, thus reducing some sources of variability in analytical methods.

4.9 INTERNATIONAL CONSIDERATIONS

Production and marketing of meat and poultry products is an enormous international business. There is an increasing need to participate in this far-ranging milieu. The actions and stakes are more complex when dealing with different regulatory policies and initiatives. Passage of the 1985 Farm Bill by the U.S. Congress required "equal to" regulatory control programs for countries exporting meat and poultry products into the United States. The directives on hormones and other residues in fresh meat which were developed by the European Economics Community (EEC) illustrate the level of interest for rules and regulations for meat and poultry products. The Codex Alimentarius Commission has sought expert advice and recommendations on issues of residues of pesticides and veterinary drugs in animals. Two Codex committees have been established to deal with these issues. Corresponding expert committees sponsored by the Food and Agriculture Organization of the United Nations (FAO) and the World Health Organization (WHO) have been established for developing acceptable daily intakes of residues of requested substances from the two Codex committees to aid in establishing international standards for these selected residues. Another principle objective for these Codex committees is the development of suitable methods of residue analysis as an aid in eliminating potential trade barriers. These actions have been justly motivated by international concerns of food safety for consumers.

This presents a dilemma for developing countries. They may be eager to participate in world markets, but find the going tough because of their limited regulatory programs.

Developing residue analytical methods is resource intensive, whether measured in terms of personnel, laboratory facilities, instrumentation, reagents or other consumable items. These considerations affect all residue control programs, but are usually more acute in economically deficient or developing countries. The strain on resources limits considerations on the types of residue control programs that can be developed, and the analytical methods employed. Countries with well-developed regulatory programs are also feeling the effects of reduced levels of resources for the quality and quantity of the testing they are called upon to produce. Cooperative efforts are called for.

Standardizing analytical methods for veterinary drug, pesticide and environmental contaminants is a worthy objective even though it may be difficult to achieve. Some of the benefits, however, include more effective resource utilization, simplified training needs, minimal duplication of methods/development, establishment of a common basis for quality assurance, and quality control programs to satisfy the criteria for international trade and to assure acceptable residue limits.

International method standardization may proceed slowly so long as inspection programs have different objectives, different good animal practices, and legal requirements. Nevertheless, standardization will be a step towards international harmony for residue control in animal food products.

4.10 SUMMARY

Agricultural production today employs a broad spectrum of pesticides as well as substances applied or administered to food-producing animals for therapeutic, prophylactic or diagnostic purposes, or for modification of physiological function and behavior.

Residue control programs began to evolve in the 1960s in response to consumer concerns about residues of environmental contaminants and agricultural chemicals such as veterinary drugs and pesticides. That evolution continues, and new programs are being developed to monitor for the generally minute quantities of residues in foods, in part, to give credible assurance to consumers that the food offered to them is as safe as present-day science and technology can make it. Analytical methods are an integral part of a residue control program; however, they are only one of the decision-making tools available to those agencies responsible for food safety and wholesomeness. As such, they complement other inspection tools.

Another important analytical function is to develop the metabolic data regulatory agencies need for drug registration. These data help establish withdrawal times and maximum residue limits. They also facilitate international trade by providing assurance that these food products meet internationally agreed-upon residue limits for food safety.

Complementary methods are needed for cost-effective regulatory control programs. Rapid tests provide qualitative to semi-quantitative data. Reliable laboratory-based quantitative methods are suitable for routine use in enforcement of maximum residue limits. They also provide corroborating residue information, including data on structure, particularly if the analytical method used is based on detection using a specific functional group of the molecule, or based on the molecule's mechanism of action. The majority of current analytical methods are quantitative laboratory methods. Confirmatory methods provide unambiguous information on structure. In some instances, quantitative data and confirmatory data are available from the same analytical procedure. The synergism of these three classes of analytical methods is important for regulatory programs because it is important to quantify and confirm what we have the ability to detect.

For regulatory control purposes, performance characteristics for analytical methods must be well-defined to insure they are properly designed, and used to yield information appropriate for regulatory decisions. Performance information is commonly available through a successfully completed multilaboratory validation study employing a minimum of three laboratories. A collaborative study consisting of six or more laboratories is preferred. Complementing method performance factors is quality assurance. Beyond assuring optimum method performance, quality assurance plans provide sound laboratory management tools to continuously monitor performance, support regulatory decisions on product disposition, and define training and education guidelines for regulatory control laboratories and analysts. Quality assurance programs promote high performance in residue analysis.

4.11 REFERENCES

Arnold, D. and A. Somogi. 1982. "Zum Nachweis von Chloramphenicol-Rueckstaenden," *Deutsche Veterinaermed Gesellschaft*, 206–212.

Burlingame, A. L., D. Maltby, D. H. Russell and P. T. Holland. 1988. "Fundamental Reviews, Mass Spectrometry," in *Anal. Chem.*, 60(12):294R–342R.

Charpentier, B. A. and M. R. Sevenants. 1988. *Supercritical Fluid Extraction and Chromatography Techniques and Applications, American Chemical Society Symposium Series No. 361,* Washington, D.C.

Codex Alimentarius Commission. 1987. Codex Committee on Residues of Veterinary Drugs in Foods, Report of the First Session of the Codex Committee on Residues of Veterinary Drugs in Foods, Washington, D.C., 27 Oct.–31 Oct., 1986, Alinorm 87/31.

Codex Committee on Residues of Veterinary Drugs in Foods. 1988. Report of the Second Session of the Codex Committee on Residues of Veterinary Drugs in Foods, Washington, D.C., 30 Nov.–4 Dec., 1987. Alinorm 89/31.

Codex Committee of Residues of Veterinary Drugs in Foods. 1989. Report of the Third Session of the Codex Committee on Residues of Veterinary Drugs in Foods, Washington, D.C., 31 Oct.–4 Nov., 1988, Alinorm, 89/31A.

Covey, T. R., D. Silvestre, M. K. Hoffman and J. D. Henion. 1987. "A GC/MS Screening, Confirmation and Quantitation Method for Estrogenic Compounds,"*Biomed. & Environ. Mass Spectrom.,* 15:45–56.

EEC Commission. 1987. "Commission Decision of 14 July 1987 Laying Down the Methods to Be Used for Detecting Residues of Substances Having Hormonal Action and of Substances Having a Thyrostatic Action," *EEL Official Journal No. L,* 87/410/EEC, 223:18–36.

FAO—Food and Agriculture Organization of the United Nations. 1985. "Residues of Veterinary Drugs in Foods, Report of a Joint FAO/WHO Expert Consultation," No. 32, Rome, 29 Oct.–5 Nov., 1984.

Henion, J. D. 1988. Unpublished Report. "Analytical Methods for the Determination of Anabolic Agents," USDA, Agricultural Research Science Contract No. 58-3615-5-038.

Horwitz, W. 1988. "Guidelines for Collaborative Study Procedure to Validate Characteristics of a Method of Analysis," *J. Assoc. Off. Anal. Chem.,* 71:160–172.

Mumma, R. O. and K. W. Hunter. 1988. "Potential of Immunoassays in Monitoring Pesticide Residues in Foods," in U.S. Congress, Office of Technology Assessment, Pesticide Residues in Food: Technologies for Detection, pp. 171–181.

Randecker, V. W., J. A. Reagen, R. E. Engel, D. L. Soderberg and J. E. McNeal. 1987. "Serum and Urine as Predictors of Sulfamethazine in Swine Muscle, Liver and Kidney," *J. Food Protect.,* 50:115–122.

Sphon, J. A. 1978. "Use of Mass Spectrometry for Confirmation of Animal Drug Residues," *J. Assoc. Off. Anal. Chem.,* 61:1247–1252.

Thomas, M. H., R. L. Epstein, R. A. Ashworth and H. Marks. 1983. "Quantitative Thin Layer Chromatography Multi-Sulfonamide Screening Procedure: Collaborative Study," *J. Assoc. Off. Anal. Chem.,* 66:884–892.

USDA. 1982. *Chemistry Quality Assurance Handbook, Volume 2.*

Youden, W. J. and E. H. Steiner. 1975. *Statistical Manual of the AOAC,* Association of Official Analytical Chemists, Arlington, VA, pp. 33-36.

CHAPTER 5

Antibiotic Residues and the Public Health

ROBERT C. WILSON, D.V.M., Ph.D.[1]

5.1 INTRODUCTION

THE evolution of antibiotic use in domestic animals followed very closely the development and therapeutic utilization of these agents in human beings. This fact is well-documented by an early textbook, *Milks' Veterinary Pharmacology, Materia Medica and Therapeutics,* published in 1949. This text gives specific indications for therapeutic use of penicillin, tyrothricin, streptomycin, streptothricin, bacitracin, chloramphenicol, clavicin and chlortetracycline. The rapid development during the 1940s of these and other antibiotics resulted in the control of many of the infectious diseases of domestic animals, and a subsequent increase in food production. This success produced an attitude that encouraged the use of antibiotics in disease therapy for food animals, and contributed to the use of these agents as feed additives. This revolution in animal care and in production was so extensive that by 1969 it was estimated that over 80% of the animals consumed in the United States were treated with one or more drugs approved by the Food and Drug Administration (Van Houweling and Kingma, 1969). There is little change in these statistics today (Parkhie, 1983; Steele and Beran, 1984).

Antibiotics are administered to food-producing animals in therapeutic or in subtherapeutic amounts. The therapeutic indications include therapy of existing disease, or prophylactically to prevent disease. Although antibiotic therapy has greatly improved livestock health it has not eliminated

---

[1]Department of Physiology and Pharmacology, Auburn University, Auburn, Alabama, USA.

the threat of disease. This fact is supported by an estimate made in 1975, that the complete control of all infectious diseases would result in an increase in food production in the United States by 10–15% (Penny, quoted by Steele and Beran, 1984). Thus, in spite of more than 30 years of antibiotic administration there remains a continued need for therapy.

The amount of antibiotics used for subtherapeutic purposes in the United States is extensive; in fact, approximately 50% (17.5 million tons) of the total production is used in this manner (Black, 1984; Yeoman, 1982; Budiansky, 1984). In the United States, subtherapeutic use is defined as the administration of less than 200 grams/ton of feed for chickens and pigs, and less than 11 mg/kg/day for cattle (Walton, 1983). This use of antibiotics is justified based upon the well-established facts that low-level feeding improves feed conversion efficiency, increases weight gain, and is an effective aid in the prevention of diseases. Initial evidence for the improved weight gain was reported during the search for the animal protein factor (APF), a factor which stimulated growth in young pigs and chickens. Although the APF was initially believed to be vitamin B_{12} alone, it was ultimately demonstrated that another substance was also involved. This unknown factor or substance was shown to be the antibiotic chlortetracycline (Ott et al., 1948; Stockstad et al., 1949; Stockstad and Jukes, 1950a, 1950b). Subsequent to this discovery, additional studies demonstrated the effectiveness of low-level feeding of antibiotics in reducing disease losses and improving feed conversion. The exact mechanisms by which these effects are produced are however, still in some dispute (U.S. Government Report, 1980).

5.2 INCIDENCE OF ANTIBIOTIC RESIDUES

The reported data on antibiotic residues found in human food must be evaluated with a thorough knowledge of the detection methodology and the accompanying sensitivity. Earlier studies, therefore, may have underestimated the actual incidence of residue, while more recent studies may be over-emphasizing the public health significance of these data. The potential overemphasis is a result of an increase in the sensitivity of the analytical methods, so that detection of drug concentrations has increased from parts per million (ppm) to parts per billion (ppb) and now, in some cases, even to parts per trillion (ppt). As the sensitivity of assays increases further, one must be cognizant of the basic pharmacokinetic principle of first-order elimination, and the prolonged elimination half-life from various tissue depots (Mercer et al., 1977; Arnold et al., 1984). If this concept is carried to its conclusion, it is conceivable that there will always be a residue following administration of a single dose of an antibiotic. The concept of zero tolerance is thus held in question, and some interpretation must be made about what constitutes a significant level of residue (Jackson, 1980; Booth, 1980; Walton, 1983; Rico, 1984). The important question is then, at what tissue concentration is the risk to humans above an acceptable level? The World Health Organization (WHO) has produced guidelines for acceptable levels of antibiotics in human food. In the United States the tolerance levels are frequently updated and published in the Code of Federal Regulations (CFR). These documents set forth the acceptable antibiotic residues in meat, liver, kidney, fat, skin, milk, various parts of the egg, and other edible tissue. Each drug approved is specific for use in each species, and for the required withdrawal time.

Surveys in search of antibiotic residues in human food have been conducted in many countries over the last several years. For example, a 1984 study performed in a Canadian red meat plant reported that only 12, or 2.5%, of 487 bovine samples tested over a 108-week period were positive for antibiotic residues (Masztis, 1984). This study was performed using the swab test on premises (STOP), a microbiological test which uses *Bacillus subtilis* as the test organism. The actual incidence of antibiotic contamination was even less, as the 487 samples (kidney or muscle) represented a selected fraction of the total number of animals slaughtered (104,951), i.e., a biased sampling program. Criteria used in selection included signs of mastitis, metritis, lameness, pneumonia, or other conditions where antibiotics may have been administered. Thus, the violative rate was only 0.011% for all the cattle slaughtered. This contrasts with an earlier study performed in 1975 where a similar microbiological assay was performed on kidney tissue obtained from slaughterhouses in Ontario and Saskatchewan (Tittiger et al., 1975). These workers found positive samples in 4 of 211 (1.9%) beef kidneys and 5 of 611 (0.8%) swine kidneys, while none of 27 sheep or 120 chicken kidneys were positive. However, when the investigators tested urine samples, the percent of samples testing positive to antibiotics increased to 3.6% of 2,108 bovine samples, 7.7% of 2,409 swine samples, and 9.7% of 176 sheep samples. This study by Tittiger suggests two facts. First, the concentration by the kidney which is demonstrated by the increased incidence of residues found in the urine, and secondly, when compared with the 1984 study by Masztis, an overall reduction in the incidence of antibacterial contamination has occurred.

Huber (1971a) reported the prevalence of antibacterial drug residues in more than 4,000 animals in the United States. He revealed the magnitude of the residue problem when he reported antibiotic residues in 27% of a group of 1,381 swine, 9% of 580 beef cattle, 17% of 788 veal calves, 21% of 328 market lambs, and 20% of 926 chickens. The antibiotics most frequently found were penicillin G, dihydrostreptomycin, tylosin, neomycin, chlortetracycline and oxytetracycline. Information gained from this and a previous study by the same investigator suggested that management and husbandry practices tended to be the controlling factor in the occurrence of residues (Huber, 1971b). Since the early 1970s, however, considerable progress has been made in reducing antibiotic residues. This is a result of both an increase in awareness of the potential for producing drug residues by the producer and of increased surveillance by the regulatory agencies. For example, antibiotic residues in swine have been reduced from an incidence of 5.7% in 1978 to an average of only 0.4% during the years 1980 to 1984 (Hall, 1986). Likewise, the incidence of antibiotic contamination in mature cattle has dropped from a violation rate of 2.2% in 1979 to a low 0.2% in 1983. A similar remarkable reduction in residues has also been reported with chicken, which has been free of residues since 1979, and a reduction in residues in turkey from 2.4% in 1979 to 0.01% in 1983 (Crawford, 1985). The decrease in residues in poultry production appears to be a direct result of improved management techniques.

Unfortunately, the reduction in antibiotic residues in red meat animals has not been as successful as in poultry. Problems continue with residues of sulfonamides in swine, and with antibacterial residues in cull dairy cows and bob veal (Parkhie, 1983; Rosenberg, 1985; Paige, 1993). Between 1973 and 1984 the percentage of swine with violative sulfonamide residues has ranged from 4.4% to a high of 13.1% in 1977 (Hall, 1986). With the total estimate of swine receiving sulfamethazine in the feed or as a therapeutic agent being as high as 80%, the incidence of these residues is not astounding. Evidence that this violative rate is continuing is confirmed by a recent suggestion by the FDA that sulfamethazine may be removed from the market due to its continued high residue rate (DVM, 1987). Likewise, as we look at the bob veal statistics, we find that in 1979 the violative rate was 7.8% for sulfonamide residues and remained at 7.7% in 1983 (Crawford, 1985). By 1988, however, significant progress was made, in that of 180,728 bob veal calves tested by the FSIS for antibiotic and sulfonamide residues using the CAST, there were only 3,095 animals positive, for a residue rate of 1.71%. The most common antimicrobial agents found in a subset of 967 CAST-positive animals were neomycin, unidentified agents, streptomycin and penicillin. Less than 1% were positive for sulfamethazine (Wilson et al., 1991).

The presence of violative antibiotic residue in milk has been controlled by existing monitoring programs and improved management (Huber, 1971a; Crawford, 1985). In fact, since the development and implementation of the specific withdrawal periods for antibacterials, few cases of violative residues have been reported (Oliver et al., 1984). The poultry and egg industry has also been effective in reducing the likelihood of drug residue to an acceptable rate.

5.3 TESTS FOR RESIDUES

Various screening tests are used to determine the presence of antibiotic residues in human food. These tests are generally microbiological in nature, and depend upon inhibition of bacterial growth (Walton, 1983). In the United States and Canada the swab test on premises (STOP) is the standard method of screening in red meat plants. This test involves insertion of a sterile swab into kidney or muscle tissue, and incubation of that swab on an agar plate seeded with *Bacillus subtilis.* A positive response is indicated by a zone of inhibition about the swab. Positive samples are confirmed in the laboratory, where testing methodology includes assay by high performance liquid chromatography, various microbiological colorimetric and radioactive tests, electrophoretic, enzyme immunoassay, mass spectrometric and others (Bories et al., 1983; Berkowitz and Webert, 1986; Bates et al., 1983; Inglis and Katz, 1978; Vilim et al., 1979; Thorogood et al., 1983; Wal et al., 1980).

5.4 SOURCES OF RESIDUES

In today's modern agricultural setting where herd and flock health are controlled and adequate records maintained, the occurrence of antibiotic residues should be unlikely. Most agricultural producers have a firm commitment to good procedures, and seek to abide by proper withdrawal periods (Meisinger and Van Houweling, 1986). Antibiotic residues that do occur in human food, in general, result from inadvertent contamination due to either human carelessness, a failure to observe label directions, or to individual variation in an animal's ability to eliminate drugs.

An analysis performed in the 1970s reported that human error was responsible for 18% of the cases of violative drug residues (Booth, 1982). This, along with other studies, suggested that errors in production, failure to properly clean feed mixing and delivery equipment, delivery of the wrong feed by the feed mill or improper storage at the farm was involved in most cases (Hall, 1986; Bevill, 1984; Mussman, 1975). In swine, the electrostatic attraction of powdered sulfamethazine to the metallic milling or storage equipment was implicated as a source of feed contamination. This problem is reduced by using granular forms of the drug, which evidently reduce the accumulation of dust (Rosenberg, 1985). Similar problems have been reported in the dairy industry, where it was estimated that the milk from a single cow treated with 200 mg of penicillin has the potential to contaminate the combined milk of up to 8,000 cows if mixed in a bulk tank (Wishart, 1983). Inadequate flushing of antibiotic contaminated discarded milk from milking equipment has resulted in violative residues in the entire bulk tank (Pugh et al., 1977; Egan and O'Connor, 1983).

Failure to abide by the approved label instructions is by far the leading cause of illegal residues found in human food. In one study it was estimated that 76% of the violative residues were the result of noncompliance with proper withdrawal times (Booth, 1982). A more recent report indicates that in 1990 46%, and in 1991 54% of illegal residues resulted from failure to adhere to proper withdrawal times (Paige, 1993). In addition, the administration of a drug at a dose above the recommended level can result in an increase in elimination half-life, or the dose-dependent pharmacokinetics. This is usually the result of some rate-limiting process in drug metabolism. Other illegal residues can occur when drugs are used by non-approved routes of administration, or given to nonapproved species (Mussman, 1975).

Even when proper label instructions are followed, various factors, such as age, disease state, and site of injection, will influence the rate of elimination of drugs from animal tissue. These individual factors can result in elimination rates different from the predicted values which are based upon pharmacokinetic data obtained from normal animals.

The most important age-related factors which help to determine the disposition and elimination of antibacterial drugs in food animals are renal function, drug distribution volume, degree of protein binding, drug metabolism and biliary excretion (Short and Clarke, 1984). The rate of development of renal function is species dependent. Several studies have indicated, for example, that while the calf may have nearly adult renal function by the second day of life (Dalton, 1968 a, b, c), it may take up to a week in the lamb (Alexander and Nixon, 1962). The elimination half-life of sulfamethazone is a good example, as the half-life is 13.5 hours in a one-day old bovine, and only 6 hours in an adult (Nouws et al., 1983). However, except for bob veal, renal function may not play a significant role in residue violations. Hepatic maturation, in contrast, is generally believed to be a two-stage process with the major development completed at 4 weeks postpartum and the second stage completed by about 10 weeks of age. Drugs which are primarily eliminated by hepatic mechanisms often demonstrate age-dependent pharmacokinetics. Chloramphenicol, a drug primarily eliminated by hepatic processes, has a half-life of 11.7 and 6.1 hours in day-old calves and pigs, and 4.4 and 0.8 hours at 8 weeks or older, respectively (Reiche et al., 1980; Svendsen, 1976). Thus, the use of drugs in young animals presents a greater potential for residue problems. Similar pharmacokinetic variation and problems could no doubt be observed as a process of aging, although most animals raised for food are slaughtered before this process has any influence on the elimination of antibiotics (Koch-Weser et al., 1982; Schmucker, 1985).

Many of the disease conditions for which antibiotics are utilized have an effect upon drug elimination, and therefore may have a bearing on drug residues present in slaughtered animals. Fever, for example, has been shown to almost double the elimination half-life of gentamicin given to febrile rabbits, and to induce change in the distribution volumes of gentamicin in horses and sheep (Halkin et al., 1981; Wilson et al., 1983; Wilson et al., 1984). The significance of these changes can be appreciated if one considers that the total body clearance of a drug is directly related to the distribution volume, and inversely related to half-life. Absorption of drugs following oral administration can be altered by fever in young goats, possibly due to inhibition of reticulo-rumen motility (Van Miert et al., 1976). Bacterial pneumonia in calves resulted in increased serum oxytetracycline concentrations, a condition which may cause prolonged elimination (Clark et al., 1974). These and other factors are no doubt involved in the persistence of violative residues following emergency slaughter, and are the reasons for the selective sampling of those animals showing signs of drug injection, mastitis, or any condition which may have required drug therapy. Drug residues may persist at the injection site for prolonged periods of time (Mussman, 1975). For example, in a study where various sulfonamides and trimethoprim were injected intramuscularly into swine, detectable residues were found at most sites 6 days after the injection, and with the sulfonamides at 30 days in almost half of the animals (Rasmussen and Svendsen, 1976). Other drugs such as dihydrostreptomycin persist for up to 60 days, while positive residues of chloramphenicol are found at 7 days post injection. Sodium and procaine penicillin, neomycin, tylosin and oxytetracycline residues have also been determined at 24 hr or more post injection (Nouws and Ziv, 1977; Nouws et al., 1986).

Although the incidence of illegal antibiotic residues in milk is infrequent, when it does occur it is often the result of some variation from the established milk-out rates. This variation can be the result of difference in the physiology and anatomy of the udder, the level of milk production, the stage of lactation, and the chemical nature of the specific drug and vehicle (Egan et al., 1981). Antibiotic residues have been found in milk following prophylactic and therapeutic use for mastitis, following intrauterine administration for treatment of metritis, or from contamination of milking equipment (Pugh et al., 1977; Oliver et al., 1984; Egan and O'Connor, 1981, 1983; Slee and Brightling, 1981; Egan and Meaney, 1985). Recent evidence suggests that some milk which tests negative for antibiotic residues may contain metabolites which are potential sources of human illness (Boonk and Van Ketal, 1982).

5.5 PUBLIC HEALTH SIGNIFICANCE OF ANTIBIOTIC DRUG RESIDUES

With the extensive use of antimicrobial agents in treatment of disease and as feed additives, it is inevitable that, due to nonconformance with withdrawal periods, or due to individual animal pharmacokinetic variation, drug residues will be found in food derived from animals. The public health significance of such adulteration of the food supply can result in several potential sequela depending upon the degree of adulteration. These potential hazards can be subdivided into four groups (Kidd, 1986; Corry et al., 1983): toxicological, microbiological, immunopathological and environmental hazards.

5.5.1 TOXICOLOGICAL HAZARDS

Most of the antimicrobials administered in therapeutic or in subtherapeutic amounts to domestic animals have been approved for use in the human. These drugs have been shown to be relatively safe and where they occur, their acute and chronic toxicities have been well-documented. In almost all cases, the potential dosage received by an individual consuming these drugs as a tissue residue will be considerably less than that obtained as a primary drug. This statement is based upon the estimated dosage that could be received by ingestion of food within a brief period of time. Single exposure to the usual antibiotic residue concentrations found in tissues other than liver and kidney would seldom be above 2 ppm (Huber et al., 1969; Mercer et al., 1971a, 1971b; Teske, 1972; Rollins et al., 1972). At this level, a person would need to ingest over 1 kg of meat to get a total dose of 2 mg of drug, a dose well below those used in therapy of primary human disease. The toxicological significance of antibiotic residues in human food therefore, must be generally related to the dose and duration of exposure. Some increase in the amount of antibiotic received would exist with excessive ingestion of specific concentrating tissue from an injection site, an unlikely event. Antibiotics which react chemically as bases, (e.g., chloramphenicol) and the macrolides (erythromycin, tylosin, and oleandomycin) are more likely to accumulate in tissue at a higher concentration than in plasma. This is due to ion trapping which results from a pH difference between blood and tissue, and to the innate lipid solubility of the compound (Knight, 1981; Wilson, 1984). A factor which could possibly reduce the potential human intake is that most animal tissue is cooked before eating, an event which will reduce the drug concentration of some agents like the penicillins and the tetracyclines. Others, such as chloramphenicol and streptomycin, appear to be somewhat thermostable (Lloyd and Mercer, 1984).

The likelihood of direct toxicity from antibiotics or their metabolites in animal tissue is extremely low, a fact confirmed by the lack of documented scientific reports in the literature (Black, 1984; Corry et al., 1983; Hewitt, 1975). An exception to this usual situation exists with chloramphenicol, a drug which produces a toxicity not related to dose (Sande and Mandell, 1985). Chloramphenicol has been implicated as the causative agent in many cases of fatal aplastic anemia, a condition which has been reported to occur with extremely low doses of this agent (Polak et al., 1972). For example, one 6-year-old girl died after receiving only 2 grams of chloramphenicol (Cone and Abelson, 1952) while a 73-year-old female died following an estimated total dose of only 82 mg received as an opthalmic drug (Fraunfelder and Bagby, 1982). That these total dosages are obtainable in human food was confirmed in a report which found chloramphenicol residues in 13 calves out of 3,020 tested (Settepani, 1984). Ten muscle samples had levels above 1 ppm while one was reported to contain 12,671 ppm. This fact led to a ban on the use of chloramphenicol in food animals in the United States.

The chronic toxicity related to antibiotic residues may be manifested by mutagenic, teratogenic, or carcinogenic potential (Kidd, 1986; Corry et al., 1983; Black, 1984). Although no reported cases are documented, the elucidation of the carcinogenic and mutagenic potential of some drugs, such as the nitrofurans, has resulted in a zero-tolerance level in human food.

5.5.2 MICROBIOLOGICAL HAZARDS

Perhaps the most important result of antibiotic use in animals is the development of multiple drug-resistant microorganisms. Because this problem is more directly related to residues, therapeutic and subtherapeutic use, than it is to drug, it will be discussed only briefly. Novick (1981) reviewed numerous reports suggesting that the subtherapeutic use of antimicrobials has both promoted and maintained the incidence of multiple drug-resistant organisms in animals raised for human consumption. Numerous studies have documented that these organisms may enter the human food chain (Walton, 1970; Shooter et al., 1970; Walton and Lewis, 1971; Linton et al., 1976; Bensink and Botham, 1983; Adetosoye, 1986). Although controversy continues about the incidence of human disease resulting from exposure to animal microorganisms, several reports give compelling evidence that it does occur (Lyons et al., 1980; Bezanson et al., 1983). Two recent studies demonstrate the potential danger to the health of the general public. In the first study, a *Salmonella newport* strain resistant to ampicillin, carbenicillin, and tetracycline, and characterized by a 38 kilobase R plasmid, was identified with an outbreak of human salmonellosis in 18 individuals from 4 Midwestern states (Holmberg et al., 1984). Before the development of clinical salmonellosis, 12 of the 18 patients had received antibiotics, primarily amoxicilin or penicillin, drugs to which the strain was resistant. A 6-state comparison of *S. newport* plasmid profiles from both human and animal isolates indicated that the original source was hamburger prepared from cattle raised in South Dakota, slaughtered in Minnesota, and processed in Nebraska. These cattle had been fed subtherapeutic amounts of chlortetracycline for growth promotion and disease prevention, a fact which may have provided the selective pressure required for maintenance of the R plasmid. In the second study, Spika et al. (1987) isolated a strain of *S. newport* from humans in California. The strain had an identical plasmid and resistance to chloramphenicol. Development of human disease, as in the previous study was often associated with prior antibiotic therapy and ingestion of hamburger. It was speculated that the prior antibiotic therapy may have given the resistant strain a competitive advantage in colonization of the human intestinal tract. This epidemic strain was isolated from hamburger, abattoirs, dairy facilities, and from sick dairy cows. Thus, the development of antibiotic resistance in animal bacteria was associated with human disease.

The possible disruption of the human gastrointestinal flora following ingestion of antibiotic residues in food has been raised (Walton, 1983). Although no direct evidence exists that alteration of the normal human flora has occurred, two potential sequela have been identified. First, the ingestion of an antibiotic may exert a selective pressure on the intestinal flora, resulting in the growth of microorganisms with acquired or natural resistance, and secondly, the direct or indirect production of resistance in pathogenic enteric bacteria. After a thorough assessment of current information, one author concluded that the risk to humans was negligible (Van Klingeren, 1986).

5.5.3 IMMUNOPATHOLOGICAL HAZARDS

Reports of allergic reactions following ingestion of antibiotic residues in food have been uncommon. The majority of the reported cases have been associated with penicillin allergy and were characterized by dermatitis (Vickers et al., 1958; Borrie and Barret, 1961; Vickers, 1964; Wicher et al., 1969; Tscheuchner, 1972). Three reactions followed consumption of milk or dairy products, or in one case after eating minced meat possibly taken from an injection site. The dose of penicillin required to produce an allergic response was small; in fact, it was estimated to be from 1 to 10 units or less (Borrie and Barrett, 1961; Tscheuchner, 1972). In another study, it is suggested that while very minute quantities of penicillin are capable of eliciting an allergic response, somewhat greater amounts are required to induce the primary sensitization (Adkinson, 1980). A more severe allergic response has been documented in one 14-year-old female who suffered anaphylaxis on four separate occasions. Before each attack, the girl, who had a positive skin test and evidence of antibodies to streptomycin, had eaten a meal containing ground beef. Although the meat was not available for testing, the individual had eaten beef before with no problem. Thus, while this case is highly suggestive of streptomycin adulteration, there was only presumptive evidence of drug contamination (Tinkelman and Bock, 1984).

The paucity of reported allergic problems due to antimicrobial drugs is surprising in light of the estimate that 10–15% of the population is allergic to these agents (Algird, 1966). This low number of reports may be due to many factors, including a failure to determine the cause of the allergic episode, as well as to insufficient amounts or absence of drug residues. A recent study on the pathogenesis of chronic urticaria in humans demonstrates the problem (Boonk and Van Ketel, 1982). In this study, 245 patients were tested and 24% (59) were sensitive to either benzyl-penicillin G or to a protein complex (penicilloyl-polylysine) which is formed when the beta lactam ring is opened by a penicillinase. Forty-two of the positive reacting patients were placed on a milk-free diet, as were 40 other patients with chronic urticaria who were not sensitive to the penicillin or its protein complex. None of these patients receiving the diets had positive scratch tests to milk. Twenty-two of the penicillin sensitive patients had good-to-excellent improvement, while only 2 of the 40 control patients had similar results. This data suggests that dairy products, even those which test negative by microbiological assay, may contain sufficient penicillin or its metabolite to maintain urticaria. Because the penicilloyl-polylysine complexes are considered to be more immunogenic than free penicillin, additional information needs to be obtained about their pharmacokinetics, and about the incidence of their residues in foods of animal origin (Adkinson, 1980).

5.5.4 ENVIRONMENTAL HAZARDS

The hazard to the public health from antibiotic introduction to the environment is not well documented. Obvious sources of antibiotics are the fecal and urine wastes (slurry) from intensive livestock production operations. It has been well documented that pathogenic organisms resistant to multiple antibiotics may persist for prolonged periods of time in this slurry (Jones, 1980), and that the resistance patterns do not change over at least a 7-week period (Hinton and Linton, 1982). Whether the resistance is maintained due to continual antibiotic exposure, or some other condition has not been established. Although no reports of human disease resulting from contact with slurry have been found, the potential reservoir of R plasmids in the environment is of concern.

Graber et al. (1979) reported the effect on the resistance of enteric *E. coli* cultured from workers spraying orchards of stone fruit with oxytetracycline. When fecal samples were cultured 2–4 months after completion of spraying, the *E. coli* from their wives and children were resistant 22.7% of the time, while the orchardists' cultures were resistant at a level of only 12.9%. These values were not significantly different, and did not markedly differ from the expected values in the general population.

5.6 SUMMARY

The presence of antibiotic residues and their significance to human health have been investigated almost from their first usage. In spite of the concern about their potential adverse effects on the general public health, little direct evidence exists which supports this fear. The reasons for the lack of substantiated reports are many, but include: inability to trace the origin of food stuff, failure to recognize adversities, failure to analyze for drug metabolites, and others. With better compliance by producers, and continued and more complete surveillance by regulatory agencies, it is possible that the future will be characterized by an even safer food supply.

5.7 REFERENCES

Adetosoye, A. I. 1986. "Carcass Contamination and Antibiotic Residue in Slaughtered Cattle at Bodija Abattoir and Slaughter Houses in Ibadan, Nigeria," *Bull. An. Hlth. Prod. Afr.*, 34:99–101.

Adkinson, N. F. 1980. "Immunological Consequences of Antimicrobials in Animal Feeds," in *The Effects on Human Health of Subtherapeutic Use of Antimicrobials in Animal Feeds,* Washington, D.C.: National Academy of Science (NAS), pp. 301–316.

Alexander, D. P. and D. A. Nixon. 1962. "Plasma Clearances of *p*-Aminohippuric Acid by the Kidney of Foetal, Neonatal and Adult Sheep," *Nature (London),* 194:483–484.

Algird, J. R. 1966. "Allergic Patient," *Conn. Med.*, 30:878–880.

Arnold, D., D. Vom Berg, A. K. Boertz, Y. Mallick and A. Somogyi. 1984. "Radiological Determination of Chloramphenicol Residues in Muscles, Milk and Eggs," *Arch. Lebensmittelhyg.*, 35:131–136.

Bates, M. L., D. G. Lindsay and D. H. Watson. 1983. "A Note on Inhibition Test and Electrophoretic Detection Limits of Antibiotics Used in British Animal Husbandry," *J. Appl. Bacterol.*, 55(55):495–498.

Bensink, J. C. and F. P. Botham. 1983. "Antibiotic Resistant Coliform Bacili, Isolated from Freshly Slaughtered Poultry and from Chilled Poultry at Retail Outlets," *Aust. Vet. J.*, 60:80–83.

Berkowitz, D. B. and D. W. Weber. 1986. "Enzyme Immunoassay-Based Survey of Prevalence of Gentamicin in Serum of Marketed Swine," *J. Assoc. Off. Anal. Chem.*, 69:437–441.

Bevill, R. F. 1984. "Factors Influencing the Occurrence of Drug Residues in Animal Tissues after the Use of Antimicrobial Agents in Animal Feeds," *JAVMA*, 185:1124–1126.

Bezanson, G. S., R. Khakhira, and E. Bollegraff. 1983. "Noscomial Outbreak Caused by Antibiotic-Resistant Strain of *Salmonella typhimurium* Acquired from Dairy Cattle," *Can. Med. Assoc. J.*, 128:426–427.

Black, W. D. 1984. "The Use of Antimicrobial Drugs in Agriculture," *Can. J. Physiol. Pharmacol.*, 62:1044–1048.

Boonk, W. J. and W. G. Van Ketel. 1982. "The Role of Penicillin in the Pathogenesis of Chronic Urticaria,"*Br. J. Derm.*, 106:183–190.

Booth, N. H. 1980. "Tissue Density of Drugs with Respect to Time," *JAVMA*, 176:1134–1140.

Booth, N. H. 1982. "Drugs and Chemical Residues in the Edible Tissues of Animals," in *Veterinary Pharmacology and Therapeutics,* 5th ed, N. H. Booth and L. E. McDonald, eds., Ames, IO: Iowa State University Press, pp. 1065–1113.

Bories, G. S. F., J. C. Peleran and J. M. Wal. 1983. "Liquid Chromatographic Determinations and Mass Spectrometric Confirmation of Chloramphenicol Residues in Animal Tissue," *J. Assoc. Off. Anal. Chem.*, 66:1521–1526.

Borrie, P and J. Barret. 1961. "Dermatitis Caused by Penicillin in Bulked Milk Supplies," *Br. Med. J.*, 2:1267–1268.

Budiansky, S. 1984. "Jumping the Smoking Gun," *Nature*, 311:407.

Clark, J. G., C. Adams, D. G. Addis, J. R. Dunbar, D. D. Hinman and G. P. Lofgreen. 1974. "Oxytetracycline Blood Serum Levels Studies in Healthy, Pneumonic and Recovered Cattle," *Vet. Med. /Sm. An. Clinician*, 69:1542–1546.

Cone, T. E and S. M. Abelson. 1952. "Aplastic Anemia Following Two Days of Chloramphenicol Therapy: Case Report of Fatality in Six-Year-Old Girl," *J. Pediatr.*, 41:340–342.

Corry, J. E. L., M. R. Sharma and M. L. Bates. 1983. "Detection of Antibiotics in Milk and Animal Tissues," in *Antibiotics, Assessment of Antimicrobial Activity and Resistance,* A. D. Russell, ed., NY: Academic Press, pp. 349–370.

Crawford, L. M. 1985. "Residues and Other FDA Concerns," *JVME*, 11:95–96.

Dalton, R. G. 1968a. "Renal Function in Neonatal Calves–Diuresis," *Br. Vet. J.*, 124:371–381.

Dalton, R. G. 1968b. "Renal Function in Neonatal Calves: Urea Clearance," *Br. Vet. J.*, 124:451–459.

Dalton, R. G. 1968c. "Renal Function in Neonatal Calves: Inulin, Thiosulphate and Para-Aminohippuric Acid Clearance," *Br. Vet. J.*, 124:498–502.

1987. "DVM:CVM Threatens Withdrawal of Sulfamethazine," from FDA Newsletter, 18:43.

Egan, J. and W. J. Meaney. 1985. "Persistence of Detectable Residues of Penicillin and Cloxacillin in Normal and Mastitic Quarters Following Intramammary Infusion," *Vet. Rec.*, 116:436–438.

Egan, J., F. O'Connor and J. Connolly. 1981. "Milk-Out Rates for Intramammary Antibiotic Preparations," *Irish J. Food Sci. Tech.*, 5:129–141.

Egan, J. and F. O'Connor. 1983. "Milking Equipment Can Be a Source of Antibiotic Residues in Milk," *Farm Food Res.*, 14:46–47.

Fraunfelder, F. T. and G. C. Bagby, Jr. 1982. "Fatal Aplastic Anemia Following Topical Administration of Ophthalmic Chloramphenicol," *Am. J. Ophthalmol.*, 93: 356–360.

Graber, C. D., S. H. Sandifer, N. H. Whitlock, C. B. Loadholt and B. J. Poore. 1979. "Acquired Resistance of Autochthonous *E. coli* in Controls and Orchardists Engaged in the Spraying of Oxytetracycline," *Bull Environ. Contam. Tox.*, 22:202–207.

Halkin, H. M. Lidji and E. Rubinstein. 1981. "The Influence of Endotoxin-Induced Pyrexia on the Pharmacokinetics of Gentamicin in the Rabbit," *J. Pharm. Exp. Ther.*, 216:415–418.

Hall, R. F. 1986. "Residue Avoidance in Pork Production," *Compend. Contin. Educ.*, 8:200–208.

Hewitt, W. L. 1975. "Clinical Implication of the Presence of Drug Residues in Food," *Fed. Proc.*, 34:202–204.

Hinton, M. and A. H. Linton. 1982. "The Survival of Multi-Antibacterial Drug Resistant *Escherichia coli* and *Salmonella typhimurium* in Stored Static Slurry from a Veal Calf Unit," *J. Hyg. Camb.*, 88:557–565.

Holmberg, S. D., M. T. Osterholm, K. A. Senger and M. L. Cohen. 1984. "Drug-Resistant Salmonella from Animals Fed Antimicrobials," *N. Eng. J. Med.*, 311:617–622.

Huber, W. G. 1971a. "The Impact of Antibiotics Drugs and Their Residues," *Adv. Vet. Sci. Comp. Med.*, 15:101–132.

Huber, W. G. 1971b. "The Public Health Hazards Associated with Non-Medical and Animal Health Usage of Antimicrobial Drugs," *J. Pure Appl. Chem.*, 21:377–388.

Huber, W. G., M. B. Carlson and M. H. Lepper. 1969. "Penicillin and Antimicrobial Residues in Domestic Animals at Slaughter," *JAVMA*, 154:1590–1595.

Inglis, J. M. and S. E. Katz. 1978. "Improved Microbiological Assay Procedures for Dihydrostreptomycin Residues in Milk and Dairy Products," *Appl. Environ. Microbiol.*, 35:517–520.

Jackson, B. A. 1980. "Safety Assessment of Drug Residues," *JAVMA*, 176:1141–1144.

Jones, P. W. 1980. "Disease Hazards Associated with Slurry Disposal," *Br. Vet. J.*, 136:529–542.

Kidd, A. R. M. 1986. "Some Principles for Establishing Withdrawal Periods," *Pig Vet. Soc. Proc.*, 16:82–87.

Knight, A. P. 1981. "Chloramphenicol Therapy in Large Animals," *JAVMA*, 178:309–310.

Koch-Weser, J., D. J. Greenblatt, E. M. Sellers and R. I. Shader. 1982. "Drug Therapy—Drug Disposition in Old Age," *N. Eng. J. Med.*, 306:1081–1088.

Linton, A. H., B. Handley, A. D. Osborne, B. G. Shaw, T. A. Roberts and W. R. Hudson. 1976. "Contamination of Pig Carcasses at Two Abattoirs by *Escherichia coli* with Special Reference to O-Serotypes and Antibiotic Resistance," *J. Appl. Bact.*, 41:89–110.

Lloyd, W. E. and H. D. Mercer. 1984. "Toxicity of Antibiotics and Sulfonamides Used in Veterinary Medicine," in *Handbook Series in Zoonoses*, Section D *Antibiotics, Sulfonamides and Public Health, Volume 1*, H. J. Jukes, H. L. Dupont and L. M. Crawford, Section eds., pp. 217–231.

Lyons, R. W., C. L. Samples, H. N. Desilva, K. A. Ross, E. M. Julian and P. J. Checko. 1980. "An Epidemic of Resistant *Salmonella* in a Nursery: Animal-to-Human Spread," *JAMA*, 243:546–547.

Masztis, P. S. 1984. "Antibiotic Residue Testing in a Beef Slaughterhouse," *Can. Vet. J.*, 25:329–330.

Meisinger, D. and C. D. Van Houweling. 1986. "Pork Producer's Points of View on Residues and Certain Zoonoses," *JAVMA*, 188:134–136

Mercer, H. D., J. D. Baggot and R. A. Sams. 1977. "Application of Pharmacokinetic Methods to the Drug Residue Profile," *J. Tox. Environ. Health*, 2:787–801.

Mercer, H. D., H. F. Righter and G. G. Carter. 1971a. "Serum Concentrations of Penicillin and Dihydrostreptomycin after Their Parenteral Administration in Swine," *JAVMA*, 159:60–65.

Mercer, H. D., L. D. Rollins, M. A. Garth and G. G. Carter. 1971b. "A Residue Study and Comparison of Penicillin and Dihydrostreptomycin Concentrations after Intramuscular and Subcutaneous Administration in Cattle," *JAVMA*, 158:776–779.

Milks, H. J. 1949. Antibiotics in *Milks' Veterinary Pharmacology, Materia Medica and Therapeutics*, Alex Eger Inc., pp 617–629.

Mussman, H. C. 1975. "Drugs and Chemical Residues in Domestic Animals," *Fed. Proc.*, 34:197–201.

Nouws, J. F. M., T. B. Vree, M. Baakman and M. Tijhuis. 1983. "Effects of Age on the Acetylation and Deacetylation Reactions of Sulfadimidine and N4-Acetyldimidine in Calves," *J. Vet. Pharmacol. Ther.*, 6:13–22.

Nouws, J. F. M, T. B. Vree, J. Holtkamp, M. Baakman, F. Driessens and P. J. M. Guelen. 1986. "Pharmacokinetic, Residue and Irritation Aspects of Chloramphenicol Sodium Succinate and a Chloramphenicol Base Formulation Following Intramuscular Administration to Ruminants," *Vet. Quar.*, 8:224–232.

Nouws, J. F. M and G. Ziv. 1977. "Persistence of Antibiotic Residues at the Intramuscular Injection Site in Dairy Cows," *Refuah. Vet.*, 34:131–135.

Novick, R. P. 1981. "The Development and Spread of Antibiotic Resistant Bacteria as a Consequence of Feeding Antibiotics to Livestock," *Ann. NY Acad. Sci.*, 368:23–59.

Oliver, S. P., R. T. Duby, R. W. Prange and J. P. Tritschler II. 1984. "Residues in Colostrum Following Antibiotic Dry Cow Therapy," *J. Dairy Sci.*, 67:3081–3084.

Ott, W. H., E. L. Rickes and T. R. Wood. 1948. "Activity of Crystalline Vitamin B_{12} for Chick Growth," *J. Biol. Chem.*, 174:1047–1048.

Paige, J. C. 1993. "Analysis of Illegal Residues in Meat and Poultry," *FDA Veterinarian*, 8:2, 5–6.

Parkhie, M. R. 1983. "Interagency Collaboration Regarding the Safety of Animal-Derived Human Food," *JAVMA*, 182:682–683.

Polak, B. C. P., H. Wesseling, A. Herxheimer and L. Meyler. 1972. "Blood Dyscrasias Attributed to Chloramphenicol," *Acta. Med. Scand.*, 192:409–414.

Pugh, K. E., P. G. Hendy and J. M. Evans. 1977. "A Possible Source of Antibiotic Contamination of Milk," *Vet. Rec.*, 101:313.

Rasmussen, F. and O. Svendsen. 1976. "Tissue Damage and Concentration at the Injection Site after Intramuscular Injection of Chemotherapeutics and Vehicles in Pigs," *Res. Vet. Sci.*, 20:55–60.

Reiche, R., M. Mulling and H. H. Frey. 1980. "Pharmacokinetics of Chloramphenicol in Calves During the First Weeks of Life," *J. Vet. Pharmacol. Ther.*, 3:95–106.

Rico, A. 1984. "Veterinary Drugs and Public Health, the General Problem of Safety," in *Safety and Quality in Food, Proceedings of a DSA Symposium*, Elsevier: Amsterdam, pp. 89–103.

Rollins, L. D., R. H. Teske, R. J. Condon and G. G. Carter. 1972. "Serum Penicillin and Dihydrostreptomycin Concentrations in Horses after Intramuscular Administration of Selected Preparations Containing These Antibiotics," *JAVMA*, 161: 490–495.

Rosenberg, M. C. 1985. "Update on the Sulfonamide Residue Problem," *JAVMA*, 187:704–705.

Sande, M. A. and G. L. Mandell. 1985. "Antimicrobial Agents, Tetracyclines, Chloramphenicol, Erythromycin, and Miscellaneous Antibacterial Agents," in *Goodman and Gilman's The Pharmacological Basis of Therapeutics*, 7th ed., A. G. Gilman, L. S. Goodman, T. W. Rall and F. Murad, eds., NY: Macmillan Publishing Co., pp. 1170–1198.

Schmucker, D. L. 1985. "Aging and Drug Disposition: An Update," *Pharmaco. Rev.*, 37:133–148.

Settepani, J. A. 1984. "The Hazard of Using Chloramphenicol in Food Animals," *JAVMA*, 184:930–931.

Shooter, R. A., E. M. Cooke, S. A. Rousseau and A. L. Breaden. 1970. "Animal Sources of Common Serotypes of *Escherichia coli* in the Food of Hospital Patients," *The Lancet*, 666:226–228.

Short, C. R. and C. R. Clarke. 1984. "Calculation of Dosage Regimens of Antimicrobial Drugs for the Neonatal Patient," *JAVMA*, 185:1088–1093.

Slee, K. J. and P. Brightling. 1981. "Antibacterial Activity of Cows Milk Following Therapy with Oxytetracycline Uterine Pessaries," *Austral. Vet. J.*, 57:143–144.

Spika, J. S., S. H. Waterman, G. W. Soo Hoo, M. E. St. Louis, R. E. Pacer, S. M. James, M. L. Bissett, L. W. Mayer, J. Y. Chiu, B. Hall, K. Green, M. E. Potter, M. L. Cohen and P. A. Blake. 1987. "Chloramphenicol-Resistant *Salmonella newport* Traced Through Hamburger to Dairy Farms," *N. Eng. J. Med.*, 316:565–570.

Stelle, J. H. and G. W. Beran. 1984. "Perspectives in the Use of Antibiotics and Sulfonamides," in *Handbook Series in Zoonoses*, Section D *Antibiotics, Sulfonamides and Public Health, Volume 1*, H. J. Jukes, H. L. Dupont and L. M. Crawford, Section eds., pp. 3–34.

Stockstad, E. L. R and T. H. Jukes. 1950a. "Growth-Promoting Effects of Aureomycin in Turkey Poults," *Poultry Sci.*, 29:611–612.

Stockstad, E. L. R. and T. H. Jukes. 1950b. "Further Observations on the 'Animal Protein Factor,' " *Proc. Soc. Exptl. Biol. Med.*, 73:523–528.

Stockstad, E. L. R., T. H. Jukes, J. Pierce, A. C. Page, Jr. and A. L. Frankline. 1949. "The Multiple Nature of the Animal Protein Factor," *J. Biol. Chem.*, 180: 647–654.

Svendsen, O. 1976. "Pharmacokinetics of Hexobarbital, Sulfadimidine and Chloramphenicol in Neonatal and Young Pigs," *Acta Vet. Cand.*, 176:1–14.

Teske, R. H., L. D. Rollins and G. G. Carter. 1972. "Penicillin and Dihydrostreptomycin Serum Concentrations after Administration in Single and Repeated Doses to Feeder Steers," *JAVMA*, 160:873–878.

Thorogood, S. A., P. D. P. Wood and G. A. Prentice. 1983. "An Evaluation of the Charm Test—A Rapid Method for the Detection of Penicillin in Milk," *J. Dairy Res.*, 50:185–191.

Tinkleman, D. G. and S. A. Bock. 1984. "Anaphylaxis Presumed to Be Caused by Beef Containing Streptomycin," *Ann. of Allergy*, 53:243–244.

Tittiger, F., B. Kingscote, B. Meldrum and M. Prior. 1975. "Survey of Antibiotic Residues in Canadian Slaughter Animals," *Can. J. Comp. Med.*, 39:178–182.

Tscheuchner, I. 1972. "Anaphylactische Reaktion auf Penicillin nach Genuss von Schweinefleisch," *Z. Haut. Gescl.*, 47:591–592.

U.S. Government Report. 1980. *Antibiotics in Animal Feed*, Serial No. 96-169, Washington: U.S. Government Printing Office.

Van Houweling, C. D. and F. J. Kingma. 1969. "The Use of Drugs in Animals Raised for Food," *JAVMA*, 155:2197–2200.

Van Klingeren, B. 1986. "Do Residues of Antimicrobial Drugs Constitute a Microbiological Risk for the Consumer?" in *Comparative Veterinary Pharmacology and Therapy*, A.S.J.P.A.M. van Miert, M. G. Bogaert and M. Debackers, eds., Norwell, MA: MTP Press, pp. 527–538.

Van Miert, A.S.J.P.A.M., H. Van Gogh and J. G. Wit. 1976. "The Influence of Pyrogen Induced Fever on Absorption of Sulpha Drugs," *Vet. Rec.*, 99:480–481.

Vickers, H. R. 1964. "Dermatological Hazards of the Presence of Penicillin in Milk," *Proc. R. Soc. Med.*, 57:1091–1092.

Vickers, H. R. and S. Alexander. 1958. Dermatitis Caused by Penicillin in Milk," *The Lancet*, 1:351–352.

Vilim, A. B., S. D. Moore and L. Larocque. 1979. "Microbiological Determination of Penicilin G, Ampicillin, and Cloxacilin in Milk," *J. Assoc. Off. Anal. Chem.*, 62:1247–1250.

Wal, J. M., J. C. Peleran and G. F. Bories. 1980. "High Performance Liquid Chromatographic Determination of Chloramphenicol in Milk," *J. Assoc. Off. Anal. Chem.*, 63:1044–1048.

Walton, J. R. 1970. "Contamination of Meat Carcasses by Antibiotic-Resistant Coliform Bacteria," *The Lancet*, 672:561–563.

Walton, J. R. 1983. "Antibiotics, Animals, Meat and Milk," *Zbl. Vet. Med. A.*, 30:81–92.

Walton, J. R. and L. E. Lewis. 1971. "Contamination of Fresh and Cooked Meats by Antibiotic Resistant Coliform Bacteria," *The Lancet*, 718:255–257.

Wicher, K., R. E. Reisman and C. E. Arbesman. 1969. "Allergic Reaction to Penicillin Present in Milk," *JAMA*, 208:143–145.

Wilson, D. J., B. B. Norman and C. E. Franti. 1991. "Detection of Antibiotic and Sulfonamide Residues in Bob Veal Calf Tissues: 967 Cases (1987–1988)," *JAVMA*, 199:759–766.

Wilson, R. C. 1984. "Microlides in Veterinary Practice," in *Macrolide Antibiotics Chemistry, Biology, and Practice*, New York: Academic Press, pp. 301–347.

Wilson, R. C., J. N. Moore and N. Eakle. 1983. "Gentamicin Pharmacokinetics in Horses Given Small Doses of *Escherichia coli* Endotoxin," *AJVR,* 44:1746–1749.

Wilson, R. C., D. D. Goetsch and T. L. Huber. 1984. "Influence of Endotoxin-Induced Fever on the Pharmacokinetics of Gentamicin in Ewes," *AJVR,* 45:2495–2497.

Wishart, D. F. 1983. "Antibiotics in Milk," *Vet. Annual,* 23:71–78.

Yeoman, G. H. 1982. "Review of Veterinary Practice in Relation to the Use of Antibiotics," in *The Control Antibiotic-Resistant Bacteria,* C. H. Stuart-Harris and D. M. Harris, eds., The Beecham Colloquia, pp. 57–75.

CHAPTER 6

Causes, Detection, and Correction of Sulfonamide Residues in Swine

RICHARD F. BEVILL, D.V.M., Ph.D.[1]

6.1 INTRODUCTION

THE sulfonamides are a class of structurally similar chemical compounds which have a bacteriostatic and antiprotozoal action. These compounds are commonly called *sulfa* drugs, and during the half-century following their introduction to treat bacterial diseases in humans, they have been used to treat diseases in swine, sheep, cattle, and poultry. Following the introduction of antibiotics (e.g., penicillin, tetracycline), the therapeutic use of sulfas in humans declined until the mid-1970s when synergistic combinations of sulfonamides and diaminopyrimidines were marketed which had 20–100 times greater antimicrobial effect than the sulfas.

Sulfonamide use in domestic animals has been sustained as a result of their relatively low cost, wide spectrum of antimicrobial activity, ease of administration in feed and water, and general effectiveness in the treatment of diseases caused by susceptible bacteria. However, in recent years, the gradual emergence of sulfonamide-resistant bacteria has narrowed the therapeutic use of sulfonamides in cattle to the treatment of pneumonia, foot rot, and acute infections of the uterus, and in swine to the treatment of pneumonias and bacterial enteritis. Daily therapeutic doses of sulfas administered to animals range from 110–220 mg/kg body weight. Therapeutic doses of sulfas are administered to cattle in the form of solid oral dose forms (boluses) or incorporated in their water supply. Swine which require

---

[1]Professor of Veterinary Pharmacology; Head, Department of Veterinary Biosciences, College of Veterinary Medicine, University of Illinois at Urbana, Illinois, USA.

therapeutic doses of sulfa are most commonly medicated by adding a water-soluble form of the drug to their drinking water.

At the present time, the low-level subtherapeutic use of antibacterial agents in animal feeds is an accepted method for increasing animal performance by preventing certain diseases, improving feed efficiency and increasing the rate of weight gain. For these purposes, sulfonamides are incorporated in swine feed at concentrations of 100 g/908 kg. Cattle rations are commonly fortified at concentrations which provide 350 mg/head/day. Most producers feed subtherapeutic concentrations of antibacterials and associate their use with profitable animal husbandry.

Since 1950 sulfas have been routinely incorporated in swine feeds to control atrophic rhinitis and Group E streptococcal infections. For these purposes and to promote growth, the sulfas are usually combined with other antibiotics. Sulfamethazine (SMZ) and sulfathiazole have been approved for continuous use in swine feeds at 100 g/ton and in various combinations with chlortetracycline and procaine penicillin or tylosin. These combinations are sold under a variety of proprietary names including Tylan-Sulfa®, ASP-250® and CSP-250®. Current estimates are that 70–80% of swine slaughtered for food in the United States have received some form of sulfa medication.

The use of certain sulfonamides has been related to an overgrowth of thyroid follicles with a tendency for tumor formation in rodents (Swarm et al., 1973). The development of dose-related thyroid gland tumors in mice fed rations containing high concentrations (4,800 mg SMZ/kg feed) of sulfamethazine over a 24-month period has been demonstrated in chronic toxicity trials (NCTR Technical Report, 1988). On the basis of this risk assessment data, the Food and Drug Administration (FDA) is considering a ban on sulfamethazine use. Continued use of sulfamethazine and other sulfonamides is possible if the drugs are shown to produce thyroid tumors by a secondary mechanism, e.g., stimulation of the pituitary to produce excesses of thyroid stimulating hormone, which predisposes to cellular overgrowth and tumor formation. Studies are currently being conducted to determine if sulfamethazine is a primary or secondary carcinogen.

When sulfonamides or other drugs are administered to animals, certain guidelines associated with safety to the animal, and the person consuming products derived from treated animals, must be followed. Guidelines for product use are included on the product label or in package inserts, and include information concerning dose, duration of dosage, and preslaughter drug withdrawal.

Withdrawal times are established in animal studies, and represent the time required for the residual concentrations of the drug in animal tissues to consistently decline below a tolerance value based on results of specific toxicity test procedures (Jackson, 1980). The current tolerance for sulfas in animal tissues is 0.1 mg/kg of tissue. When swine were slaughtered at sequential times following the withdrawal of a ration containing 100 gm SMZ/ton, and resupplied with a nonmedicated ration, the concentration of the drug in tissues was equal to or less than 0.1 mg/kg within 7 days in all tissues sampled (Messersmith et al., 1967). The current withdrawal time for sulfamethazine is 15 days and for sulfathiazole 7 days. These withdrawal times are more than twice as long as those predicted by slaughter studies designed to establish the minimum time required for sulfas in tissues to be below the value.

Despite the two-fold safety factor included in the withdrawal time for sulfamethazine, violative residues (those exceeding tolerance) have been detected in pigs sampled at slaughter during the past 15 years. In fact during this period, sulfamethazine has produced more violative drug residues than any other drug administered to animals. From 1973 to 1978, 10–15% of the swine slaughtered for food contained violative residues (Brown, 1979). However, the incidence was reduced to less than 5% during the mid-1980s, and, with the introduction of rapid testing methods, declined to less than 2% of the animals slaughtered during the past 2 to 3 years. Although the incidence of violative sulfonamide residues has been greatest in swine, they have also been detected in veal calves, poultry, and cow's milk.

The frequent occurrence of violative sulfonamide residues in swine was first reported by the USDA's Food Safety and Inspection Service (FSIS) in 1974 after a National Residue Program had been established to provide for monitoring and surveillance of drug residues (Engel, 1980). Although the incidence of all violative drug residues in animals occurred at a low frequency (1 to 2% of samples tested) violative sulfamethazine residues were present in 9.6% of the samples which tested positive.

The purpose of this chapter is to discuss in detail the causes of violative sulfa residues in swine, and methods for detection, prevention, and correction of the problem.

6.2 CAUSES OF SULFONAMIDE RESIDUES

6.2.1 LOW-LEVEL CONTAMINATION OF FEEDS WITH SULFONAMIDES

Some producers first realized that sulfonamide residues were a problem in 1977. At that time, FSIS chemists were determining the sulfa content of 1,200 tissue samples a month, and FSIS and the FDA had asked producers to voluntarily reduce the occurrence of sulfa residues by strictly observing withdrawal time. Despite the efforts of producers to reduce the incidences of violative sulfa residues, they continued to occur.

The residue situation intensified as FSIS identified herds which were the source of violative residues, and suspended marketing privileges of the producers until they were able to demonstrate the absence of sulfonamides in a five-animal test group marketed from their premises. Subsequent investigations as to the cause of violative sulfa residues in swine have provided an expanded perspective on the epidemiology and prevention of drug residues in food-producing animal species.

FDA records assembled during 1970–1977 indicated that failure to observe withdrawal time, delivery of the wrong rations by feed mils, inadequate cleaning of bins and use of drugs in species for which they had not been approved were the probable causes of drug residues. Failure to observe withdrawal time was listed as the probable cause of residues in 76% of the cases investigated (Booth, 1982).

Many producers whose marketing privileges had been suspended as a result of marketing swine with violative residues continued to insist that proper preslaughter drug withdrawal procedures had been observed on their farms. Some indicated that no sulfas were being used on their farms. The accuracy of FSIS testing methods, the original withdrawal period for sulfamethazine and the importance of residues ranging from 0.1 to 1.0 mg/g of tissue as a hazard to humans were questioned by swine producers and commercial interests (Black, 1978).

Laboratory groups working closely with swine producers to determine the source of sulfonamides responsible for violative residues soon concluded that in some instances residues were being caused by factors other than the producer's failure to withdraw the drug prior to slaughter. One group at the University of Illinois which was completing contractural research for FDA concerning the absorption, distribution, metabolism, and excretion of sulfamethazine in swine became involved in the residue problem as a result of requests from swine producers seeking solutions to their residue problems. The possibility that extremely low concentrations of sulfamethazine in feed could produce residues was explored after standard pharmacokinetic calculations indicated that concentrations as low as 2.0 mg SMZ/kg of feed could produce blood concentrations of the drug which exceeded tolerance. Subsequent feeding trials in which animals were fed rations containing 2.0, 4.0, and 8.0 mg SMZ/kg produced violative sulfamethazine residues in swine liver if fed during the preslaughter drug withdrawal period (Bevill, 1978; Ashworth and Epstein, 1986). These results indicated that as little as 40 lbs of feed containing 100 mg SMZ/kg could contaminate a ton of unmedicated finishing-withdrawal feed with sufficient drug to cause violative residues in swine if the feed was consumed during the preslaughter drug withdrawal interval.

When the disposition of sulfamethazine and sulfathiazole administered intravenously to swine was compared, the half-life for elimination of sulfamethazine was 12 hours (Limpoka, 1978). The corresponding value for sulfathiazole was 1.2 hours (Koritz et al., 1978). The 10-fold shorter half-life of sulfathiazole decreases the tendency for the drug to accumulate in the body as reflected by its shorter withdrawal period (7 days) and lower blood and tissue concentrations. Drugs like sulfathiazole and sulfamethazine are eliminated by first-order kinetic processes, and as a rule-of-thumb, the amount of time required for 99.99% of the drug to be eliminated from the body following dosing is equivalent to nine half-lives (10.8 hours for sulfathiazole and 108 hours for sulfamethazine). When a constant amount of sulfa is infused intravenously or absorbed from the intestine, we find that per unit time the drug will accumulate in blood and tissues until a steady-state concentration is achieved. The time required to achieve 99.99% of steady-state value is equivalent to 9 times the elimination half-life. Therefore, sulfamethazine continues to accumulate in plasma for 108 hours before it reaches its maximum concentration at steady-state, whereas sulfathiazole will reach its maximum but much lower steady-state concentration in 10.8 hours. This explains why equivalent doses of sulfamethazine accumulate to higher concentrations in plasma and tissues than sulfathiazole and why trace amounts of sulfamethazine persist in tissues for longer times.

The administration of sulfamethazine in feed provides a continuous source of drug for absorption from the intestine. If the feed contains 1 to 2 mg SMZ/kg, the maximum steady-state concentration of the drug in blood and liver will exceed tolerance after the animal has consumed the feed for 108 hours. These blood and tissue concentrations will be maintained as long as the animal consumes the ration. Concentrations of sulfathiazole 8 times greater (16 g/kg) would be required to produce violative residues in liver when fed during the preslaughter drug withdrawal interval. Thus, differences in elimination half-life between sulfamethazine and sulfathiazole, and its effect on drug accumulation and elimination provide a valid explanation for the much higher incidence of violative sulfamethazine residues in swine.

During 1978 and 1979, USDA scientists and swine producers conducted a series of cooperative field studies to establish the importance of low-level sulfonamide contamination of rations as a cause of violative tissue residues in swine (Brown, 1979). In one study, finishing-withdrawal rations were collected on two types of farms—those producing residue-free swine and those producing animals with residues. Sulfamethazine residues equal to or exceeding 2 mg/kg were found in 14% of the samples collected. Sulfonamides were also detected in individual ration components, such as soybean meal and vitamin premixes. This initial study indicated that low-level sulfa contamination of feeds was a principal cause of violative sulfa residues in swine.

Studies designed to determine the extent and causes of sulfamethazine carryover in feeds were conducted in 65 commercial batch mills (Teague et al., 1980). These studies provided evidence which indicated that the concentration of carryover of sulfamethazine in corn flushed through the mill following preparation of a feed medicated with the drug was not influenced by the type of mixer, mixer discharge, conveying system, electrostatic grounding, or the addition of liquid to the medicated feed. The study indicated that it was very difficult to produce feed not contaminated with sulfa in batchtype plants if sulfa-medicated feeds were prepared in the unit.

6.2.2 INABILITY TO DETECT LOW-LEVEL SULFA CONTAMINATION OF FEEDS

Although low-level contamination of feeds with sulfamethazine could produce violative residues, many producers received reports from commercial laboratories indicating that their feeds contained no sulfonamides. These feed samples were commonly submitted by producers whose marketing privileges had been suspended as a result of shipping animals found to contain violative residues when slaughtered. Subsequent examination of the sensitivity of the "official" assay method revealed that concentrations less than 5 mg SMZ/kg of feed were not detected. The fact that 1 to 2 mg SMZ/kg feed would produce violative residues in swine when fed during the preslaughter withdrawal interval, coupled with an official detection method incapable of detecting those concentrations in feed, provided a logical explanation of why feed found to contain no sulfa when assayed produced violative residues when fed. The development and use of a more sensitive feed assay method revealed that many finishing-withdrawal feeds were contaminated with concentrations of sulfamethazine sufficient to produce residues when fed to finishing swine immediately prior to slaughter.

6.2.3 PHYSICAL CHARACTERISTICS OF SULFONAMIDES

The sulfonamides have an electrostatic character, and tend to cling to the inside of mixers, augers, elevator legs, and other components of feed preparation equipment used on the farm and in commercial units. This property appears to be involved in sulfa carryover between batches of medicated and nonmedicated feeds. In order to decrease the potential for electrostatic accumulation of sulfamethazine in feed production equipment, FDA insists that granulated sulfamethazine products be used to prepare sulfa-medicated feeds. However, the results of research completed at the University of Kentucky with granulated and nongranulated sulfa would indicate that in typical farm-type feed production units, the amount of dust carryover in a two-ton vertical, twin-screw mixer and a one-ton horizontal paddle mixer are of minor consequence as a contaminant (Cromwell et al., 1982). The researchers found a maximum concentration of 269 g of sulfamethazine/ton in dust removed from the mixer after a one-ton batch of medicated feed had been mixed. However, the dust weighed only 3.3 kg, and the total sulfamethazine in the dust was only 0.9 g. This amount of sulfamethazine carryover would be insufficient to produce a carryover concentration in the next 908-kg batch of feed produced in the mixer if no other sources of contamination were available. The work completed at Kentucky clearly indicates that the feed remaining in the mixer is a more important source of carryover contamination than dust. The carryover concentration in feed was essentially the same when granular and nongranular sulfamethazine products were used. However, one study conducted in Indiana indicated that concentrations of sulfamethazine in finishing feed was four times higher when powdered sulfamethazine was used to prepare rations rather than granular sulfamethazine. Although not stated, the higher carryover associated with the powdered sulfamethazine (a concentrated product) was probably due to the unauthorized use of excessive amounts of the powdered drug, which increases the chance for cross contamination. In some instances, producers have introduced the granulated products into their rations through the grinder access port of their mix-mill, which reconverts the granulated product to a fine dust within the feed preparation system, and completely negates the benefits associated with granulation.

6.2.4 FAILURE TO OBSERVE PROPER SULFAMETHAZINE WITHDRAWAL TIMES OR USE OF EXCESSIVE CONCENTRATIONS OF SULFAMETHAZINE IN FEEDS

Some users of sulfamethazine simply ignore instructions included on product labels which detail proper preslaughter withdrawal of the product from their finishing rations. In one study conducted in Minnesota, 7.3% of swine producers indicated that they did not take withdrawal times seriously. In trials conducted by an Indiana veterinarian in 1987, more than 50% of the violations detected were caused by a failure to observe the 15-day preslaughter withdrawal. Current testing procedures can be used to quickly identify swine producers who fail to observe withdrawal times, since the magnitude of sulfa residues in the tissues of animals originating on their farms will generally be much higher than those caused by sulfonamide carryover between medicated and nonmedicated rations.

In some instances, negligence and accidental contamination of feed stuffs with sulfamethazine contribute to the residue problem. Because of the constant hazard of feed contamination associated with the use of sulfamethazine (2% contamination will produce feed capable of causing violative residues), producers must exercise extreme care during the mixing process. Trained, alert individuals must be given adequate time to properly observe all precautions associated with the production of finishing-withdrawal rations containing sulfamethazine. Accidental placement of rations containing sulfamethazine in feeders or bulk tanks during the finishing-withdrawal phase has caused violative residues in swine. An incident of this type was documented by a survey team seeking sources of sulfamethazine contamination of a farm from which animals with violative residues originated. In this instance, a grower ration containing 100 g SMZ/908 kg was placed in a bulk tank which provided a feed source for swine during their finishing-withdrawal phase of production. Nonmedicated feed containing an improper protein concentration for growing swine had been placed in an adjacent bin usually reserved for sulfa-medicated rations. No identifying labels or numbers had been applied to either bulk tank.

In a study conducted by Purdue University animal scientists 18% of 80 pork producers included in the survey were routinely using in excess of 100 g/908 kg (legal limit in feeds) in their rations. The use of increased concentrations of the drug greatly increases the chances for occurrence of violative residues, since the amount of carryover feed required for critical contamination is decreased in direct proportion to the amount of excess sulfa added. As an example, if 18.2 kg of carryover feed containing 100 g/908 kg would provide critical contamination ($\geq$ 2 g/908 kg) of 908 kg of nonmedicated feed, then an equivalent level of contamination could occur if 4.54 kg of medicated feed containing 400 g SMZ/908 kg remained in the mixer following discharge of the medicated ration. It should be evident that the use of excessive unapproved concentrations of SMZ in rations increases the risk of critical cross-contamination of finishing-withdrawal feeds and greatly reduces the design flexibility of feed production and conveying system with respect to residual feed carryover.

6.2.5 RECYCLING OF SULFAS IN ANIMAL WASTES

Some producers found that violative residues continued to occur in swine produced on their farms in spite of the fact that they were observing prescribed withdrawal times, and had assured themselves that their finishing feeds during the withdrawal period contained no sulfamethazine or sulfathiazole. In one instance, a five-animal market test group of pigs fed only shelled corn during the preslaughter drug withdrawal period was found to have violative residues when slaughtered. Field investigations by researchers at the University of Illinois detected a high concentration of sulfamethazine in dried sawdust used to bed the animals. Further inquiries revealed that the bedding, which was in a seldom-used building, had been contaminated several years previously when rations containing sulfamethazine had been fed to animals confined there. Blood samples were collected from a test group of animals fed shelled corn but with access to the contaminated bedding, and analyzed for sulfamethazine content. Concentrations exceeding 0.2 mg SMZ/L blood indicated that the animals had violative tissue residues. Blood concentrations of sulfamethazine were not detected 5 days after the animals were prevented access to the area with contaminated bedding. A pilot study was subsequently conducted in which groups of animals, confined in pens with concrete floors which retained some urine and feces, were fed rations containing 100 gm sulfamethazine/908 kg, or no medication for two weeks (Bevill and Biehl, 1979). Some animals were transferred to unmedicated manure packs, some to medicated manure packs, and some animals were held in the pens in which they were fed. Blood samples collected from the animals indicated that drug-free animals transferred to sulfa-contaminated manure packs became contaminated within 72 hours by consuming urine and feces containing sulfamethazine. Sulfamethazine disappeared from the blood of animals fed the medicated ration at a slower rate when the animals were held in pens with medicated manure packs than when the animals were transferred to pens containing noncontaminated urine and feces. These findings were later confirmed by other investigators (Whipple et al., 1980). Investigators have also noted the cyclic occurrence of violative drug residues in animals confined in Cargill-type confinement units during the winter months. These units have a slanted floor exposed to the elements. Open-front housing units are located at the high end of the floor and a waste collection area formed by a concrete backslope is located at the lower end of the floor. A fence usually prevents animals access to the collected waste. In some units, the collected waste becomes so extensive that it extends into the area to which the pigs have access. If sulfas are fed during the finishing period, they accumulate in the wastes. If the animals have access to these wastes during the preslaughter withdrawal period, they may consume and absorb enough sulfamethazine to produce violative residues in their tissues at slaughter. During the winter months, these collection areas frequently freeze and the animals can no longer browse the wastes. However, when the weather warms,the wastes become liquid and are again consumed by the animals. The occurrence of violative sulfa residues in market animals raised on these floors can often be associated with periods of warm winter weather when liquid wastes containing sulfamethazine are available to the animals during the preslaughter withdrawal interval. Violative residues do not occur when the wastes are frozen.

Systems for confinement rearing of swine are often designed in such a manner that intermittent water flows are used to flush urine and feces from waste gutters, which are located at the low end of a sloped feeding area. In some instances, the animals are separated from the wastes by concrete slats or metal floor grates. In many units, the animals are given free access to the waste gutter where they are especially prone to congregate during the water flushing cycle. Water used to flush the wastes is most often directed to a system of anaerobic waste storage lagoons where solids precipitate. Later, water free of suspended solids is recycled back through the waste gutters. This recycled water may contain sulfamethazine if the drug is fed to animals whose urine and feces are flushed into the lagoon system. Sulfamethazine seems to be very resistant to degradation in lagoons. In one study utilizing artificial lagoons containing various combinations of water, dirt, feces, and sulfamethazine, 36.4% of the drug originally introduced into the artificial lagoons was still present after 6 months incubation in a chamber exposed to natural weather and light conditions (Bevill and Biehl, 1979). Sulfamethazine concentrations of 8 to 9 mg/L have been found in some farm lagoons. These concentrations may be sufficient to cause violative sulfa residues at slaughter if consumed by animals during the preslaughter drug withdrawal interval. No studies have been reported which determined the actual concentrations of sulfamethazine in recycled wastewater which would produce violative tissue residues if consumed during the preslaughter drug withdrawal interval.

Although violative residues have been traced to sulfamethazine acquired through pig contact with feces, urine, or recycled wastewater, the actual percentage of animals contaminated in this manner appears to be low when compared with the animals contaminated as a result of low-level contamination of feed with sulfamethazine.

6.3 CHARACTERISTICS OF FEED HANDLING OR WATER SYSTEMS WHICH PREDISPOSE TO CARRYOVER OF SULFAMETHAZINE FROM MEDICATED TO NONMEDICATED RATIONS

6.3.1 GRINDER-MIXERS

Most grinder-mixer units used by producers to prepare feeds on the farm are either portable or stationary types. Physical reduction in the size of grains used to prepare rations is usually accomplished with hammermills or roller mills. The ground grains are usually discharged into a mixer which may be of vertical-screw or horizontal-paddle design.

Vertical-screw mixers consist of a cylindrical tank with a cone-shaped bottom in which the vertical screw or mixing auger is centrally located. The auger is usually surrounded by a tube to facilitate the movement of the contents from the bottom to the top of the mixing chamber. Entries for the addition of concentrates, minerals, vitamins, and drugs are usually in the form of short horizontal augers which transport these materials to the mixing chamber. The mixed product is discharged to bulk tanks and feeders by way of discharge augers of varying length. Material to be discharged is usually taken from the bottom of the cone, and variable amounts of residual material are left in the equipment after the discharge augers have completed emptying the mixer chamber. Points where medicated feed or dust may accumulate include the bottom of the mixer cone and in the tubes of the feed additive and discharge augers. Well-designed vertical-screw grinder-mixer units should be equipped with clean-out ports located near the bottom of the mixing chamber. These ports should be readily accessible and easily opened to facilitate removal of residual medicated feedstuffs. The feed ingredient hoppers and augers are areas where feed additives may accumulate, and subsequently flake off and contaminate nonmedicated rations.

Horizontal-paddle mixers are less frequently used on the farm, but offer the advantage of being more readily accessible than most vertical-screw mixers. In most instances, one or more exit ports are located in the floor of the half-cylindrical mixing chamber. The mixer can be accessed from the top and cleaned in a fairly thorough manner with long-handled, soft-bristled brooms and/or low-pressure streams of air.

6.3.2 FEED CONVEYORS

A variety of transport devices, including screw augers, pneumatic tubes, and bucket elevators, may be used to move feed from the site of preparation to sites of storage and/or consumption.

Screw augers are constructed with a gap between the outermost edges of the screw flights and the surrounding tube-like housing. This gap is required to prevent wear due to contact between the metal screw flights and housing. Because of the gap, a layer of feed is allowed to remain on the floor of the conveyor tube after feed is no longer present at the entry end of the auger. This residual feed, which may contain medication, is removed when the next batch of feed is conveyed through the auger tube. The amount of feed which can remain in a screw auger depends on the size of the gap and the length of the auger. In some systems, augers may be connected at drop-boxes which retain variable amounts of residual feed.

It is difficult to insure that pneumatic-tube feed delivery systems contain no residual feeds. If medicated and nonmedicated feeds are used in the system, cross-contamination is likely to occur unless specific precautions are observed.

6.3.3 FEED STORAGE UNITS

Prepared feeds are frequently conveyed to and stored in vertical bins which may be square with a slanted bottom converging on a discharge port, or cylindrical with a cone-shaped bottom. Normally, a column of feed located directly above the bottom discharge port is the first to leave the bin when the discharge port is opened. Materials located at the top of the bin then flow toward the center, and an inverted cone of feed is formed. The bin empties from the top down, and materials placed in the bin first are frequently the last to be removed. Also, there is a tendency for feeds to adhere to the sidewalls of the bin, particularly where the vertical bin sides join the sloping bin bottom. The tendency of feeds to resist lateral movement and to accumulate at these sites may be due to the use of bins with improper bottom slope. A minimum of 60° slope on the bin bottom is required for the most ground materials. If feeds adhere to bin sidewalls, or if the bin is refilled before it is completely emptied, small amounts of residual feed may break off and enter the flow of material moving downward through the central flow path. If the residual feed contains medications, they may become mixed with clean feeds added to the bin and produce erratic contamination of nonmedicated rations.

6.3.4 FEEDERS AND FEED SYSTEMS

A variety of devices ranging from simple to complex may be used to provide rations to swine. In some systems, rations are intermittently dropped onto the pen floor where the feed is consumed. These systems have a low potential for carryover since residual feed is minimized. However, feed may build up in pen corners if the animals are overfed, and this material may serve as a source of sulfamethazine contamination during the finishing-withdrawal period.

In most production units, feeds are delivered by auger or pneumatic conveyor system to square or cylindrical feeders. These feeders utilize gravity and swine-activated agitator rods located in troughs at the bottom of the feeder to supply small amounts of feed as needed. Depending on the shape and construction of the feeder, a considerable quantity of a medicated feed could remain in a feeder that appears to be empty.

In cylindrical feeders with slanted bottoms, feeds have a tendency to adhere at the junction of the vertical sidewalls and the feeder bottom. Many of these cylindrical feeders have a cone in the bottom of the feeder to evenly divert feed to a trough from which the animals eat. Feed lodged on the diverter, or around the outside edges of the circular feeding trough, may serve as a source of medicated feed carryover when unmedicated finishing-withdrawal rations are added to the feeder.

Locations in square self-feeding devices where feeds tend to be sequestered include areas at the bottom of the feed-holding bin and under the folded metal edges used to shape the tops of the troughs. Feed frequently hardens on the bottoms and corners of the feed troughs which require special attention when shifting from medicated to finishing-withdrawal rations.

6.3.5 BULK DELIVERY TRUCKS AND AUGER DELIVERY WAGONS

Bulk delivery trucks may be used to deliver a commercially-prepared complete ration, concentrates, or other ration components to a farm. When rations fortified with sulfamethazine are delivered in the truck, contamination of subsequent feed or concentrate deliveries with sulfamethazine may occur. The degree of contamination hazard which actually exists depends on the amount of residual medicated feed which may be sequestered in the bulk delivery tank or unloading augers. Most sequestration occurs as a result of feed bridging in the tank, and feed remaining in the gaps between the auger flighting and tube. In one Indiana study, Purdue researchers found rations containing sulfamethazine which was traced to soybean meal that the producer purchased and which was very likely contaminated during delivery to the farm.

6.4 DETECTION OF SULFONAMIDE RESIDUES

In order to establish the occurrence of sulfonamide residues in a swine herd, accurate quantitative assays with a limit of detection of 10–20 ng/mL of serum, 100 ng/mL urine, and 0.1–0.2 mg/kg of feed are required. Experimental studies have verified that a predictable relationship exists between the concentration of sulfamethazine in blood or urine and the tissue of swine (Randecker et al., 1987). Concentrations of sulfamethazine in kidney, liver, and muscle tissues will equal or exceed a tolerance of 0.1 mg SMZ/kg tissue when coexisting concentrations in serum are approximately 0.19, 0.11 and 0.42 mg/L, respectively. The corresponding values in urine are 0.63, 0.37, and 1.3 mg/L. Although no exact relationships have been established to relate the concentration of the drug in feed to concentrations in various tissues, a rule-of-thumb approximation of 0.1 mg SMZ/L of serum for each mg SMZ/kg of feed being consumed at the time the serum is collected from the animal can be used as a general estimator. These relationships between serum, urine, and feed concentrations of sulfamethazine have been used to establish screening systems whereby veterinarians, producers, or regulatory officials can predict with reasonable certainty the concentrations of sulfamethazine in swine tissues prior to slaughter. In addition, FSIS personnel can screen animals prior to slaughter by collecting urine samples, and after slaughter by collecting tissue samples from the animals. Two assays are currently used to check for the presence of sulfonamides in biological fluids and feedstuffs.

The sulfa-on-site (SOS) test uses thin layers of silica gel developed with organic solvents to separate individual sulfas from each other and substances normally present in urine which could interfere with subsequent visualization of the sulfa. The test is a semiquantitative procedure which allows a trained person to estimate liver and muscle tissue sulfamethazine residues using urine as a test medium. To perform the test, high and low standards containing known concentrations of sulfa compounds and urine are spotted in separate migration lanes which have been prescored in the silica layer on the plate. The applied samples are dried with warm air, predeveloped in methanol, dried, and developed in ethyl acetate. Following development, the plates are dried and sprayed with fluorescamine, a reagent which converts the sulfa to a fluorescent derivative. The formation of the fluorescent sulfa derivative permits very small amounts of the drug to be visualized as a bright, yellow-gold fluorescent band on the fluorescamine-treated plate when viewed under ultraviolet light. The distance which the fluorescent bands move from the point of sample application is used to identify sulfas. Band position in lanes to which urine samples are applied are compared to band position in the lanes containing standard solutions. Equal movement of bands derived from standards and samples permits identification of the sulfa in the sample. The approximate concentration of sulfonamides present in urine samples is determined by comparing the relative brightness of the flourescent band in the standard and sample lanes. When the sample bands have a similar or greater intensity than standards, sulfa residues are present. The test is calibrated so that fluorescent intensities can be used to predict whether liver or muscle exceed tolerance. The SOS test was developed by FSIS for use by food inspectors or veterinary medical officers to test swine urine collected during slaughtering operations. Recently it has been used on the farm to test for the presence of sulfamethazine in animals being shipped for slaughter. Collection of urine from representative animals in a slaughter group is best accomplished early in the day when animals are aroused from a night's sleep. Sample collection is facilitated if a wire hoop is attached to the end of a broom handle or light metal tubing. A paper cup can be inserted in the loop. As urine is voided by different animals, the samples are collected individually in clean paper cups to prevent dilution. Commercial laboratories and some veterinarians have purchased the equipment and supplies required for performing the SOS test, and have been trained in its use. In Illinois, producer groups have used the SOS test as a major component of a sulfa-certification pilot project to avoid violative sulfa residues in slaughter swine. An important objective of this study is to warn the producer of a potential residue problem before the animals are sent to slaughter. If on-farm tests or in-plant tests detect a sulfa contamination problem, an approved, accredited veterinarian who has completed a residue detection program and been trained in proper use of the SOS test will review the on-farm situation and recommend specific corrective measures. To enter such a program, the producer must agree to engage in an initial and continuous testing of market animals and/or sows to maintain a certified herd status. These types of programs will continue, since FSIS will expedite the development of premarketing testing programs for detecting swine with violative sulfa residues and encourage SOS pretesting of live animals by producers.

The EZ-Screen Sulfamethazine® is a screening test which can be used to detect the presence of the drug in urine, serum, feed, water and tissue. The test produces a visible color change which indicates whether the sample contains sulfamethazine at or greater than a designated concentration. It is less expensive than other tests and can be performed by producers in less than five minutes. The test is fairly easy to run but is very sensitive. It detects concentrations of sulfamethazine of 5–10 ng/L in urine and may indicate hogs are positive for sulfamethazine when they are not in violation. Illinois trials comparing the results of the EZ-Screen serum test and SOS urine test indicated that the EZ-Screen tests had many false positives and false negatives, and were accurate slightly more than 50% of the time.

Other thin-layer chromatographic procedures of high-pressure liquid chromatographic procedures are acceptable methods of testing for the presence of sulfamethazine in serum, urine, or feedstuffs. These tests may be performed by state-funded diagnostic laboratories or privately owned analytical laboratories. The major disadvantage associated with the use of the testing facilities is the time interval required for shipment and analysis of samples. Laboratories with quick turnaround times and reasonable fees may provide a suitable alternative to on-the-farm tests in some instances.

6.5 METHODS TO PREVENT OR REDUCE THE OCCURRENCE OF SULFA RESIDUES

The primary objective of any program designed to eliminate or reduce the occurrence of violative sulfonamide residues in swine is to provide pigs with a sulfa-free environment during the 15-day preslaughter withdrawal interval. Sulfa-safe checklists provide routine operating guidelines by which producers can establish and maintain a sulfa-free environment during this interval. These checklists instruct producers to:

- Read labels attached to prepared feeds or prepackaged ration components to determine recommended dose and withdrawal time.

- Avoid overdosing or simultaneous use of sulfas in feed and water.
- Use only approved products containing sulfas for feeding purposes. It is illegal for the producer to add powdered forms of sulfas to feed. These concentrated forms of sulfas increase the chance for residues since they increase the chance for carryover, mistakes in formulation, and excessive concentrations of sulfa in portions of the ration due to nonuniform mixing.
- Establish a system for proper storage, identification, and inventory of all medications. An up-to-date record of medications received and dispensed should be maintained. These records can be used to trace the source of accidental contamination when rations are misformulated. Storage rooms should be vacuumed frequently to reduce the amount of airborne medication and medicated feed residue.
- Observe drug withdrawal times before marketing animals. Failure to withdraw rations containing sulfas at prescribed times prior to slaughter guarantees the presence of violative residues at slaughter, and possible exposure of consumers to a sulfa-adulterated product.
- Sequence the preparation of medicated and nonmedicated rations, and flush feed preparation and milling equipment following the medicated feed preparation sequence. It is recommended that a volume of ground corn or medicated feed equivalent to at least 5% of the mixer capacity be used as a flush. These techniques help to reduce cross-contamination of feeds fed during the preslaughter withdrawal interval.
- Place nonmedicated withdrawal feeds in clean bins, and deliver to clean feeders with clean distribution equipment.
- Save samples of feed fed during the withdrawal period (farm-prepared or commercial), and record which animals received the feed. Hold samples until animals have been marketed. These samples may be useful in tracing the source of contamination should violations occur.
- Identify animals receiving preslaughter withdrawal rations, and keep them separated from wastes generated by medicated animals. To reduce the possibility for urine or fecal recycling of sulfas, clean the pens daily after sulfas are removed from feed and water.

These recommendations, when practiced by responsible individuals, will provide a sulfa-safe environment for finishing swine. All owners and employees should be well-versed in the use of these procedures, and checklists should be posted in appropriate locations throughout the production unit. When possible, óne well-trained, responsible individual should be assigned the task of preparing animal rations to assure that uniform feed preparation procedures are in use at all times. Sulfa concentrations in urine or serum samples collected from swine consuming finishing-withdrawal rations for at least 10 days prior to sampling provide an excellent method for estimating the animal's exposure to sulfonamides during the preslaughter medication withdrawal interval. This is the critical production interval—when the concentrations of sulfonamides established in tissue during other phases of the production interval should decline to values less than the established tolerance of 0.1 mg/kg of tissue. In most instances, surveys, whether they are conducted by the producer for purposes of self-evaluation, or by veterinarians or regulatory personnel for purposes of certification in sulfa-safe programs, or to determine the cause of violative residues in market swine, should be initiated by collecting blood or urine samples from market hogs consuming sulfa-free rations for 10 days prior to sampling. Urine or serum samples should also be collected from sows if they are receiving rations containing sulfamethazine, or if sow feeds are used to flush mixers used to prepare sulfa-medicated rations. The concentration of sulfonamide in these samples serves as a guide for planting subsequent phases of the evaluation. Although rations fed during the withdrawal period need not be sampled at this time, it is recommended that a few representative samples be obtained and analyzed to facilitate the evaluation process should excessive concentrations of sulfas be detected in plasma or urine. Initial feed samples should be collected from sources of actual consumption (e.g., feeder troughs, feed drops in floor-type feed systems).

If the serum or urine samples contain no sulfonamides, or concentrations which indicate that the animals will be marketed with tissue concentrations less than established tolerances, then the producer should continue to use procedures outlined in sulfa-safe checklists and schedule resampling at 4–6 month intervals to check for continued compliance. Plasma samples containing 0.1 mg SMZ/L or less, and urine samples containing less than 0.4 mg SMZ/L but greater than the limits of detection of the analytical method, indicate that no violative residues will be present in the animal at slaughter if the animals are not exposed to feeds or animal wastes containing sulfas after sampling. The SOS test utilizes sulfamethazine standards in urine of 0.4 and 1.3 mg/L which correspond to a 50% finding of at least 0.1 mg/kg in liver and tissue, respectively. Feed samples collected concurrently with these plasma or urine samples should contain 1 mg SMZ/kg or less. Serum concentrations of 10 mg/L, and urine concentrations of 30 mg/L or greater, will be present when sulfamethazine withdrawal is not practiced, and the animals continue to consume rations containing 100 g SMZ/908 kg until slaughtered. The magnitude of sulfamethazine concentration in urine and plasma thus serves as an excellent guide to the source of contamination. Concentrations in the range of 0.1–0.5 mg SMZ/L in plasma or 0.4–2.0 mg SMZ/L in urine are indicative of low-level exposure to sulfamethazine, and may be the result of cross-contamination in feed, or recycled sulfa originating in animal wastes on pen floors or lagoons.

If urine and plasma samples are found to contain concentrations of sulfamethazine in excess of those considered safe, the movement of animals to market should be interrupted until the source of contamination can be identified and corrected.

Since contaminated rations represent the most common source of violative residues in animals, the feeds being consumed by the animals should be sampled and assayed for sulfas. As previously noted, concentrations in feed exceeding 1–2 mg SMZ/kg will produce violative residues if the animals are constantly exposed during the 15-day preslaughter drug withdrawal interval. Intermittent exposure to these low drug concentrations may or may not produce violative residues if the exposure is short-term and occurs near the beginning of the drug withdrawal interval. If the results of the feed assays range from 1–10 mg SMX/kg, then cross-contamination of finishing-withdrawal feed by carryover from rations containing sulfamethazine should be suspected, and the feed preparation, storage, and distribution systems should be evaluated. This evaluation should include the methods used by the operator for fortifying the ration with sulfamethazine as well as techniques for mixing, flushing, and sequencing rations.

A thorough visual examination of the feed system immediately after a feed-mixing, storage, and delivery cycle has been completed is the first step in evaluating the potential for carryover. Grinders, mixers, feed augers, feed storage units, and feeders should be examined for sites where residual feed remains in the system. The most likely areas where feeds accumulate have been discussed previously in this chapter. Areas in the system which should receive special attention include clean-out ports on grinder-mixers or mixers, points of entry of medicated premixes, augers, drop boxes, or feed retention in storage bins and feeders.

The system should be designed so that routine cleaning of areas which retain feed can be quickly and easily accomplished. Otherwise, on busy days producers will fail to clean the units properly when shifting from medicated to nonmedicated rations. Long augers and augers connected with drop boxes in which medicated and nonmedicated feeds are sequentially transported are impossible to clean and very difficult to flush. When augers of this type are incorporated in a distribution system, the operator is encouraged to rearrange the system so that separate augers can be used to convey either medicated of nonmedicated feed. Be sure that bins used to store medicated and nonmedicated feeds are labelled and that transfer systems used to convey feeds from the mixers to medicated and nonmedicated storage have no common components which cannot be thoroughly cleaned. If the same feeders are used to deliver medicated and nonmedicated feeds, be sure that they are thoroughly cleaned before nonmedicated finishing withdrawal feeds are added. A wet-dry vacuum with a large suction canister and long suction hose simplifies removal of residual medicated feed from feeders. If medicated feed is caked in the troughs, inverting and washing the feeder with a high-pressure cleaning unit is justified.

After a thorough visual evaluation has been completed, a checklist for cleaning residual feed from the system should be prepared and posted in the feed preparation area. The checklist should be one which can be routinely followed by the operator when changing from medicated to nonmedicated feed preparation and distribution. Then, a batch of feed medicated with 100 gm SMZ/908 kg should be ground, mixed, stored and distributed to animals requiring medicated feed. The operator should then remove residual feeds from the system and prepare a batch of nonmedicated feed. Samples of this nonmedicated flush should be taken at specific points in the system, and at the beginning, middle, and end of the preparation and delivery process. Samples should be collected at the mixer discharge, at the end of auger systems used to convey feeds to storage bins or feeders. After these samples are analyzed, the capability of the system to mix and deliver a sulfa-free ration without an intermediate flush can be evaluated. In most instances, the initial samples collected from the system will contain sufficient sulfamethazine to produce contamination if consumed by swine during the preslaughter drug withdrawal interval. Samples collected at the midpoint of the batch may or may not contain critical concentrations of the drug, but samples collected at the end of the delivery process should be free of sulfamethazine. If the final samples of flush collected at storage bin or feeder entries contain in excess of 1 mg SMZ/kg, then the points of lingering contamination should be identified. In most instances, the problems will be localized in distribution augers. In such cases, the system should be modified to eliminate the problem areas – if this is not possible the owner should be encouraged to either discontinue the use of the drug or develop separate systems for the preparation and delivery of medicated and nonmedicated feeds. In some instances, the use of an auger transport wagon to deliver feed to remote locations may be substituted for long, difficult-to-clean augers.

If the final samples of flush contain less than 1 mg SMZ/kg, then the operator should be encouraged to sequence the preparation of medicated and nonmedicated feeds to minimize unit cleanout and flushing. Sequencing feed preparation so that several batches of nonmedicated feed can be prepared after medicated feeds have been removed from the mill by cleanout and flushing should be encouraged, since it greatly reduces the labor and time required to produce sulfa-free withdrawal rations, and thus encourages operators to adhere to operating checklists. Sequencing also reduces the amounts of feed required to flush the system, which can be a major problem in some units. In many units, the nonmedicated flush or flushes may consist of sow rations to which sulfamethazine is not normally added. After medicated feeds are prepared and distributed, the mill is cleaned and flushed with one or more batches of the sow feed prior to preparation of the nonmedicated withdrawal rations, and the chance for sulfa carryover in the withdrawal feed is greatly reduced. However, one must remember that when sows are fed these rations, they must be fed a sulfa-free ration for 15 days prior to slaughter. On some farms where all feeder pigs are purchased and sulfamethazine is included in rations from purchase until the preslaughter drug withdrawal interval, producers may not have nonmedicated feed available to flush equipment. In these instances, a holding tank designed to hold 1–2 tons of ground shelled corn may be installed to provide a holding area for nonmedicated ground shelled corn used as a flush. This material can be subsequently used to prepare a medicated ration. Avoiding the use of sulfamethazine in pigs weighing more than 125 lbs will also provide a group of finishing animals which can be fed the nonmedicated rations used to flush the system during the time interval required for the animals to reach premarket weights when finishing-withdrawal rations would be fed. Some producers have chosen to have medicated rations prepared at commercial mills and placed in on-farm storage tanks reserved for medicated feeds. The farm mixing unit can then be used to prepare only nonmedicated rations. Other producers have simply decided to avoid the risk of sulfamethazine carryover and discontinued use of the drug. However, producers who discontinue use of the drug should collect samples of premixes, soybean meal, and other ration components delivered to the farm, and maintain a feed sample storage bank until animals receiving feeds prepared with these substances have been marketed.

If no sulfamethazine is detected in the rations being consumed by animals voiding urine containing sulfamethazine in sufficient concentrations to be predictive of violative residues at slaughter, the possibility of urinary or fecal recycling of sulfamethazine as a drug source must be considered. To insure that recycling of wastes does not occur, animals receiving sulfa-free rations should be isolated from animals being fed medicated rations. When medicated and nonmedicated animals are maintained in adjacent pens, feces and urine voided by the medicated animals may be expelled through openings in pen dividers into the pens where nonmedicated animals are kept. After medicated animals are placed on nonmedicated feeds in preparation for marketing, the floors of the pens in which they are housed should be thoroughly cleaned and then flushed with water on three successive days. This allows the major fraction of sulfamethazine present in bowel contents at the time of medicated feed withdrawal to be absorbed and voided in urine or expelled in the feces and flushed from the pens during the period of maximum drug excretion. Any contact of pigs with recycled lagoon wastes should be prevented during the preslaughter drug withdrawal interval.

Contamination of water supplies with sulfamethazine has been reported. This includes central source contamination (pond or well) as well as contamination of water lines and tank type waterers. Water samples should be collected from supply devices in the animals' pens and assayed for sulfamethazine content whenever unexplained residues occur in swine, and when feed-borne contamination has been eliminated as a possible source. There have been reports that water lines contaminated with sulfamethazine during administration of medication can be a persistent problem. Residues present in water tanks used to supply sulfamethazine for the treatment of acute diseases are frequently overlooked as sources of contamination. These tanks should be thoroughly cleaned immediately following the discontinuation of therapeutic sulfa use.

6.6 SUMMARY

Producers should recognize that the use of sulfamethazine in swine rations requires daily vigilance on the part of the farm operator and his suppliers to prevent the occurrence of violative sulfa residues in swine. The use of the drug is accompanied by a persistent residue hazard, which can be economically damaging to the producer as a result of condemnation, indirect losses suffered during suspension of marketing privileges and reduced consumer acceptance of pork.

6.7 REFERENCES

Ashworth, R. B. and R. L. Epstein. 1986. "Sulfamethazine Blood/Tissue Correlation Study in Swine," *Amer. J. Vet. Res.*, 47:2596–2603.

Bevill, R. F. 1978. "Application of Pharmacokinetics to the Study of Sulfonamide Behavior in Cattle, Sheep and Swine," *Proceedings 1st Symposium of Veterinary Pharmacology and Therapeutics*, D. R. Short, ed., pp. 75–101.

Bevill, R. F. and L. G. Biehl. 1979. "Fecal Recycling and Other Factors Contributing to the Incidence of Sulfa Residues in Swine," *Seminar Proceedings of American Pork Congress*, pp. 42–43.

Black, N. 1978. "Producers Blast Sulfa Program," *National Hog Farmer*, 23:146.

Booth, N. H. 1982. "Drug and Chemical Residues in the Edible Tissues of Animals," in *Veterinary Pharmacology and Therapeutics,* 5th ed., N. H. Booth and L. E. McDonald, eds., Ames, IO: Iowa State University Press, Chapter 66.

Brown, R. I. 1979. "Results of USDA-Producer Cooperative Field Study on the Sulfa Problem," *Seminar Proceedings of American Pork Congress,* pp. 44–46.

Cromwell, G. L., R. L. Hutagalung and R. S. Stahly. 1982. "Effects of Form of Sulfamethazine (Powder vs. Granular) on Sulfa Carryover in Swine Feeds," *J. Anim. Sci.,* 55(Supp 1):267.

Engel, R. E. 1980. "Current Food Safety and Quality Service Residue Control Program," *JAVMA,* 176:1145–1147.

Jackson, G. S. 1980. "Safety Assessment of Drug Residues," *JAVMA,* 176:1141–1144.

Koritz, G. D., R. F. Bevill, D. W. A. Bourne and L. W. Dittert. 1978. "Disposition of Sulfonamides in Food Producing Animals: Pharmacokinetics of Sulfathiazole in Swine," *Am. J. Vet. Res.,* 39:481-484.

Limpoka, M. 1978. "The Disposition of Sulfamethazine in Swine," Master's Dissertation, University of Illinois.

Messersmith, R. E., B. Sass, H. Berger and G. O. Gale. 1967. "Safety and Tissue Residue Evaluations in Swine Fed Rations Containing Chlortetracycline, Sulfamethazine and Penicillin," *JAVMA,* 151:719-724.

National Center for Toxicology Research. 1988. "Chronic Toxicity and Carcinogenicity Studies of Sulfamethazine in B6C3F$_1$ Mice," Report No. 418, Jefferson, Ak.

Randecker, V. W., J. A. Reagan, R. E. Engel, D. L. Soderberg and J. E. McNeal. 1987. "Serum and Urine as Predictors of Sulfamethazine Levels in Swine Muscle, Liver and Kidney," *Journal of Food Protection,* 50:115-122.

Swarm, R. L., G. K. S. Roberts, P. C. Levy and L. R. Hines. 1973. "Observations on the Thyroid Gland in Rats Following the Administration of Sulfamethoxazole and Trimethoprim," *Tox. and Appl. Pharm.,* 24:351-363.

Teague, H. S., L. T. Frobish and R. J. Gerrits. 1980. "USDA Sponsored Feed Mill Sulfamethazine Carryover Study," USDA-FDA Swine Industry Task Force Report on Sulfonamide Residue "Self-Help" Program, Washington.

Whipple, D. M., G. Samuelson, G. E. Heath and D. H. Showalter. 1980. "Tissue Residue Depletion and Recycling of Sulfamethazine in Swine," *JAVMA,* 176:1348-1352.

CHAPTER 7

Hormones

DON A. FRANCO, D.V.M., M.P.H., Dipl. ACVPM[1]
CATHERINE E. ADAMS, Ph.D., R.D.[2]

7.1 INTRODUCTION

THE early medical literature defined *hormones* as substances secreted by various endocrine glands and transported in the bloodstream to the target organ on which their effect was produced. The definition has been modified and applied to various substances not produced by special glands but having similar action. Nonetheless, this chapter will limit discussion of the hormones within the spectrum of anabolic agents with sex hormone-like activities. Emphasis will be placed on the use of anabolics in the U.S., the major public health issues, and a brief analysis of the international perspective.

Anabolic agents have been used in livestock agricultural practice for over 30 years to increase growth rate, reduce feed required for growth, and for tissue redistribution from fat to lean (USDA, 1987). These effects, which vary from 10–40%, can be achieved predominantly in ruminants. The use in other food animals is of little importance at the present time. The preparations available vary as to type of intrinsic hormonal activity, and the expression of the anabolic effects varies within the subpopulations of the target animal species (Hoffmann and Evers, 1986). The predominant use was in cattle and calves, and to a far lesser degree in sheep. Some experimental use to evaluate efficacy and applicability to other livestock

---

[1]Director, Scientific Services, National Renderers Association, and Director, Scientific Affairs Animal Protein Producers Industry, Alexandria, Virginia, USA.
[2]Director, Quality Control Systems, Campbell Soup Company, Camden, New Jersey, USA.

classes (e.g., swine) and poultry has been conducted, but the research trials were neither exhaustive nor comprehensive.

Ever since the initial use of these agents, and the findings that implicated diethylstilbestrol (DES) with carcinogenesis that resulted in a subsequent ban of its use in the U.S. in 1978, a universal concern evolved on the potential hazard to public health of anabolic residues in food-producing animals (Lu and Rendel, 1976).

In spite of the persistent concerns relative to harmful effects on consumers of meat from treated animals, either with naturally occurring sex steroids or synthetic hormones, the U.S. Food and Drug Administration (FDA) has determined that the approved anabolics, when used according to label directions, are safe and that no physiological effect could be expected by consuming the animal tissue. The only potential risk to public health – a remote risk – might result from continuous consumption of meat from implantation sites where classical type implants are used, resulting in the persistence of a relatively large supply of hormone. The current policy of having the implantation site of anabolic implants limited to parts automatically discarded at slaughter, e.g., the ear, should preclude this possibility.

7.2 CLASSIFICATION

Anabolic agents can be classified according to their chemical structure or origin as endogenous sex steroids, steroidal compounds not occurring naturally, and nonsteroidal compounds not occurring naturally (Hoffman and Ever, 1986). Anabolics can also be classified in three estrogenic classes: natural estrogens (e.g., 17β estradiol), phytoestrogens (e.g., zeranol, a derivative of zeralonenone, a mycotoxin), and completely synthetic estrogen, as exemplified by the now banned diethylstilbestrol (DES) (Maghuin-Rogister, 1987).

Another classification scheme puts the agents into four categories: steroid hormones, synthetic hormones, xenobiotic compounds, and other growth promoting agents (GPA's) (USDA, 1987). Another acceptable classification approach is: natural steroids (e.g., estradiol), artificial steroids (e.g., trenbolone) and nonsteroid compounds (e.g., zeranol).

For the purpose of uniformity, a description of the categories of the anabolic agents will be presented to review the characteristics and provide examples under different classes.

7.2.1 STEROID HORMONES

Steroid hormones include the estrogenic (female) and androgenic (male) hormones. They are normally produced naturally in nearly all vertebrates, especially in the gonads, the adrenals and the placenta. Testosterone formation can also take place via peripheral conversion of certain precursors like androstenedione. During pregnancy, the placenta of most species becomes the main source for estrogens. These hormones generate sexual characteristics, maintain reproduction, and stimulate growth. They are also responsible for the differences in growth and maturity between the sexes. Three of the five anabolic agents approved for use in the U.S. are steroidal hormones – estradiol, testosterone, and progesterone. The chemical structure of natural steroidal hormones is the same in many species (Fowler, 1984).

7.2.2 SYNTHETIC HORMONES

Synthetic hormones, such as diethylstilbestrol (DES), hexosterol, and melengestrol acetate (MGA), are analogues of endogenous steroid hormones, and produce effects similar to steroidal hormones. They are also similar in molecular structure to steroidal hormones, but are technically not steroids. Some of these compounds can have adverse effects on treated animals, and leave residues in meat and excreta. That is the reason why, with the exception of melengestrol acetate (MGA), the others were not approved or were once approved then subsequently banned. DES was banned for use in the U.S. on July 6, 1979.

Trenbolone acetate (TBA) is a widely used synthetic steroid with anabolic properties administered to cattle either alone or in combination with estradiol-17β or zeranol as a subcutaneous ear implant to improve weight gain, feed conversion, and nitrogen retention in many countries of the world (WHO, 1988).

7.2.3 XENOBIOTIC HORMONES

Xenobiotic hormones are derived from plants, and may not have a molecular structure related to steroidal hormones. Zeranol is the best known and most widely used compound of this class of hormones. It is a derivative of resorcyclic acid lactone, was first identified with a common parasitic corn mold, called zearalone, and was later developed as the anabolic agent, zeranol. It is a weak estrogen, and evidence indicates it acts mainly through pituitary gland stimulation, increasing the production of growth hormone, prolactin and cortisol (USDA, 1987). Zeranol is used extensively in cattle by subcutaneous implant in the ear to improve weight gain and feed efficiency. It is also used in sheep in some countries (WHO, 1988).

In cattle, the rate of depletion of the implant peaks at 5–15 days and then slows with time. Approximately 60% of the initial dose remains at the implant site after 65 days. When good animal husbandry practices were followed, the maximum mean residue levels of zeranol, calculated as zeranol equivalents, did not exceed 0.2 μg/kg in muscle, 10 μg/kg in liver, 2 μg/kg in kidney, and 0.3 μg/kg in fat at any period after implementation (WHO, 1988). In cattle, zeranol elimination occurs predominantly by way of the feces and urine.

7.2.4 OTHER GROWTH PROMOTING AGENTS (GPA's)

Other growth promoting agents (GPA's) have been studied in recent years, many demonstrating dramatic and efficacious results, and a potential positive impact on livestock production. Some of these emerging growth promoting compounds include substances such as somatotropins [growth hormones (GH)] which are naturally produced hormones that regulate growth and metabolic processes, growth hormone releasing factors (GHRF), and somatostatins.

There is prevailing enthusiasm in many sectors that some of these new compounds would have a revolutionary effect on the future of livestock agriculture in this and other countries that are prepared to embrace the concepts of applicability and recommended use. This sentiment is not consensus, however, because there is active opposition to the ready acceptance of these theories. In reality, opponents are convinced that these type substances have the potential to be a serious environmental/public health problem. They cite serious ethical issues, and genetic dangers associated with approval of these new compounds. The emotionally charged debate continues among the supporters for GPA's use and the opponents who demand more research and scientific validation before any consideration for approval can be granted. Thus, the substances are not yet approved for use in livestock production.

In summary, research reveals that hormones and other growth promotants are effective. For example, bovine somatotropin (BST) increases milk production in dairy cows, and porcine somatotropin (PST) increases feed efficiency in hogs, diminishes the amount of fat, and increases muscle size with less total feed intake. GH-releasing factors and somatostatins regulate the production of somatropins. Their approval for future use remains uncertain, and, doubtless, the proponents and opponents will continue debate within both the scientific and public arenas.

Table 7.1 lists anabolic agents that have FDA approval for use with various species.

7.3 THE USE, ACTION AND EFFECTS OF ANABOLIC AGENTS IN FOOD-PRODUCING ANIMALS

For over thirty years, anabolic agents have been used worldwide in cattle and sheep. They increase the growth of the animal, improve feed efficiency, and enhance protein depositions/carcass composition by affecting

TABLE 7.1. Anabolic Agents: U.S. Food and Drug Administration Approval and Animal Class Designation.

Product	Manufacturer	Major Chemical Component	US-FDA Approval Date	US-FDA Approved Animal Use			
				Calves	Steers	Heifers	Lambs
COMPUDOSE	Elanco	Estradiol-17β	3/12/82	Yes	Yes	Yes	
FINAPLIX	Hoechst	Trenbolone acetate	6/17/87		Yes	Yes	
HEIFER-OID	Ivy Laboratories	Testosterone propionate & estradiol benzoate	7/24/84	Yes	Yes	Yes	
MGA	Upjohn	Melengestrol acetate	6/3/77	Yes	Yes	Yes	
RALGRO	International Minerals & Chemical	Zeranol	11/5/69	Yes	Yes	Yes	Yes
STEER-OID	Ivy Laboratories	Progesterone & estradiol benzoate	11/12/82		Yes		
SYNOVEX C	Syntex	estradiol-17β & progesterone	4/9/84	Yes			
SYNOVEX H	Syntex	Testosterone propionate & estradiol benzoate	7/16/58			Yes	
SYNOVEX S	Syntex	Progesterone & estradiol benzoate	2/20/56		Yes		

the metabolic process of the animal (Michel and Baulieu, 1983). This latter function takes place by exerting a positive influence on the intermediary metabolism, either by increasing anabolic processes or inhibiting catabolism (Byers and Schelling, 1986). As a result of this process, weight gain is usually increased, and this gain is accomplished with an overall decrease in feed consumption (USDA, 1987). Thus, the anabolics increase overall rates of carcass and total protein accretion along with yield of lean, and ultimately contribute to consumer demand for a leaner beef and a move away from fat cattle (Present, 1987).

Cattle have greater feed requirements per pound of weight gain than do hogs or poultry. The use of anabolic agents has been clearly validated, both through extensive research findings and through animal husbandry observations that weight gain can be improved through proper use of anabolics. Generally, there is a wide range of effects on beef production. Experience indicates that good management and animal husbandry practices, genetics, environmental hygiene and sanitation contribute to the gains that varied widely from a 5–20% improvement in weight, 5–12% in feed efficiency, and a very impressive lean meat enhancement of 15–25%.

Additionally, beef tenderness, juiciness, and flavor are not adversely affected by anabolic use in steers and heifers. Steers show a tendency to gain weight more rapidly after anabolic use, thus creating a potential for slaughter at an earlier age, and a greater margin of profit to producers. For example, feedlot steers have approximately 12% growth promotion and 9% feed efficiency claims, while heifers' claims are approximately 10% and 9% respectively. Nonimplanted steers would require 12% more days on feed to reach the same slaughter weight as an implanted steer, while consuming 9% more feed. Additionally—and very significant to producers, the beef trade, and consumers—is that the implanted animal has better carcass composition, with a greater lean-to-fat tissue ratio. Stated differently, the nonimplanted animal has more subcutaneous fat than the implanted animal (USDA, 1987).

Byers and Schelling (1986) summarized this topic appropriately:

> Research has provided new insights into mechanisms by which growth promoting implants modify growth in beef cattle. Protein growth is a daily function, like interest in a bank, and cellular mechanisms establish the maximal rates for daily protein synthesis. In actuality, cellular limits for protein growth are not often reached due to physiological factors including hormonal and nutritional mechanisms which set priorities for and limits to protein deposition. Cattle of different types have different priorities on protein growth at different rates of growth, and larger, mature-size cattle direct more energy toward protein growth at any rate of growth. These priorities for protein growth are enhanced by anabolic implants by redirecting nutrients to protein from fat in a "daily double play," increasing lean growth at the expense of fat . . . especially at rapid rates of gain.

7.4 PUBLIC HEALTH IMPACT/INTERNATIONAL DEBATE

The use of anabolic agents in beef production has been scrutinized by research scientists for approximately twenty years. Investigations intensified in the latter part of 1970 because of possible health effects in man from consuming beef from animals that were fed or implanted with diethylstilbestrol (DES). The international scientific community evaluated the public health implications, and DES was subsequently banned for use in livestock throughout the world. This regulatory response is close to a perfect example of consensus in scientific surveillance, and represents a hallmark in the interrelationship of science and law to positively protect the public health (Crawford, 1988a). It is important to remember that DES was shown to be carcinogenic in laboratory animals prior to any confirmatory evidence that the chemical was also carcinogenic in humans (Rall, 1986).

Speculation for the emotional response and public apprehension relative to the use of hormones in man or animals today originated with DES, and the negative health consequences that the drug had on women after therapeutic use. The chemical initially used in the U.S. between the 1940s and 1950s was administered to over 2 million pregnant women for threatened abortion. Follow-up findings of the female offspring who were exposed to DES while in utero showed that nearly 80% developed glandular epithelium resembling that of the endometrium—and presumably of Muellerian origin—in the vagina or cervix, known as adenosis. Additionally, a small proportion of women with adenosis developed clear cell adenocarcinomas of the vagina, and, to a lesser degree, the cervix when in their teens or twenties (Herbst et al., 1971).

As a sequel to the banning of DES in 1978, some European factions wanted an assessment of other anabolic agents in use. Reservations about safety persisted, and a consumer activist consortium convinced public officials to initiate an extensive evaluation by scientists and the drug industry. This intense study of the approved anabolics (estradiol-17β, progesterone, testosterone, zeranol, and trenbolone acetate) was useful in proving the safety of the compounds. Various research findings validated the safety of the use of these products when used as directed. This validation was reflected in subsequent reports of the Joint Expert Committee on Food Additives (JECFA) and the Lamming Committee of Experts (Crawford, 1988b). In spite of the prevailing scientific consensus, use of anabolic agents in livestock production was banned in European Economic Community (EEC) countries as of January 1, 1988—a very unilateral and disappointing decision, it was felt, by those in the livestock industry. A report by the European Toxicology Forum summarized the practical problems the EEC hormone ban presents. In reality, the legislation is close to impossible to enforce, and the illegal use of the banned hormones and hormone "cocktails" has been documented throughout the Community. Producers are convinced of the attributive effect and safety of properly used hormones, and some are prepared to take risks and continue use through an apparently flourishing black market trade.

The prevailing opinion in the U.S. is that public health is not compromised by the use of approved hormones in animals (cattle and sheep). This is exemplified by the rigorous approval process based on scientific evidence administered by the Food and Drug Administration, and a careful residue control program managed by the Food Safety and Inspection Service. It is significant for the purpose of debate to note that minute quantities of the three natural hormones (estrogen, testosterone, and progesterone) are present in all mammals, including humans. Adjunctly, the approved hormones administered according to directions will not increase the level of that hormone more than 1% in the animal body, and 0% to consumers of meat from treated animals. In fact, bull beef in its natural form contains concentrations of the hormone testosterone that are much higher than in meat from steers or heifers that have been treated with hormone implants (Crawford, 1988b).

Unlike the endogenous (natural) sex steroids, humans and animals do not produce synthetic hormones, and thus the Food and Drug Administration requires extensive toxicological and metabolic testing and proof of safety, i.e., proof that residues deplete below the concentration considered safe. These studies are generally conducted using rodents, and provide the needed information for regulatory analyses. Toxicity, pharmacological effect, teratogenicity, dosage, cellular morphology, oncogenicity, serologic response and other parameters are all required assays in the evaluation process. Thus, the approved synthetics (trenbolone acetate and zeranol) have been subjected to a logical, rational, and scientifically defensible regulatory approach to insure public health protection. The synthetic growth hormones exhibit identical metabolic responses and results in food-producing animals as the natural hormone compounds they mimic.

7.5 SUMMARY

Public health protection is paramount in the drug approval process. Natural and synthetic hormones are subjected to rigorous analytical testing before approval for use in food-producing animals. Proper distribution and use of approved drugs is useful in animal production. Consumer safety is assured by establishing safe residue levels in meat based on stringent criteria and the use of scientific methodologies, surveillance and research. The threat to public health lies in improper and illegal hormone use, including circumvention of manufacturers' recommendations. The evidence clearly indicates that meat from animals correctly treated with anabolic agents is safe for human consumption. Hormone use, however, continues to be plagued by periods of sensationalism and misinformation that confuse and prolong debate about their use. The real challenge is the widespread dissemination of information that will educate producers, the consuming public, and policymakers about both the benefits and possible health hazards of hormone use in animal production. And, above all, manufacturers must produce a product that will be safe for:

(1) The animal to which it will be administered
(2) The environment
(3) The people handling the product
(4) Consumers who will eat food from treated animals

The subject of residues in foods will continue to be of great interest to consumers who demand reassurance that drugs are used in ways to ensure safe residue levels in edible animal products. Consumerism will dictate the future demands on research endeavors to address the perceived concerns to maintain a safe food supply.

7.6 REFERENCES

Byers, F. M and G. T. Schelling. 1986. "Integrated Growth Management for Production of Beef Lean," *Symposium at Syntex Animal Health, Inc.,* Scottsdale, AZ.

Crawford, L. M. 1988a. "The Scientific and Nonscientific Aspects of the Hormone Issue," Presented at *European Meeting of the Toxicology Forum, International Agency for Research on Cancer,* Lyon, France. Unpublished.

Crawford, L. M. 1988b. "The United States Views on the Use of Hormones for Growth Promotion in Food Animals," Presentation before the Board of Inquiry of the European Parliament, Strasbourg, France. Unpublished.

Fowler, D. E. 1984. "The Economic Potential of Growth-Promoting Agents in Beef," Discussion Paper No. 84, Cantebury, New Zealand: Agricultural Economics Research Unit, Lincoln College, pp. 3–4.

Herbst, A. L., H. Ulfelder and D. C. Poskanzer. 1971. "Adenocarcinoma of the Vagina: Association of Maternal Stilbestrol Therapy with Tumor Appearance in Young Women," *N. Engl. J. Med.,* 284:878–881.

Hoffman, B. and P. Evers. 1986. "Anabolic Agents with Sex Hormone-Like Activities: Problems of Residues," in *Drug Residues in Animals,* A. G. Rico, ed., Orlando, FL: Academic Press Inc., pp. 111–146.

Lu, F. C. and J. Rendel. 1976. In *Environmental Quality and Safety, Anabolic Agents in Animal Production,* Suppl. Vol. V, F. Coulston and F. Corte, eds., Stuttgart: Thieme.

Maghuin-Rogister, G. C. 1987. "Anabolics," Conference Paper, Journees Internationales Societe Francaise de Toxicologie.

Michel, G. and E. E. Baulieu. 1983. "Mode of Action of Anabolics," In *Anabolics in Animal Production: Public Health Aspects, Analytical Methods and Regulations,* E. Mossonier, ed., Symposium held at Office International des Epizooties, pp. 53–64.

Preston, R. L. 1987. "Role of Anabolic and Repartitioning Agents in the Production of Lean Beef," *Southwest Nutrition and Management Conference.* Unpublished.

Rall, D. P. 1986. "Relevance of Results from Laboratory Animal Toxicology Studies," In *Public Health and Preventive Medicine,* J. M. Last, ed., Norwalk, CN: Appleton-Century-Crofts, pp. 515–518.

USDA, Food Safety and Inspection Service. 1987. "Economic Impact of the European Economic Community's Ban in Anabolic Implants," USDA, FSIS, Planning Office, pp. 1–20.

World Health Organization. 1988. "Evaluation of Certain Veterinary Drug Residues in Food," *Technical Report Series 763,* Geneva: WHO, pp. 26–27.

CHAPTER 8

Miscellaneous Growth Promotants

JUDI WEISSINGER, D.V.M., Ph.D.[1]

8.1 INTRODUCTION

WHERE feed costs represent the majority of the cost of producing cattle, feed additives that promote improvements in feed efficiency and weight gain are important to rapid attainment of market weight (Olentine, 1982). In addition to hormones and low level antibiotics, ionophores, traditionally used for their anticoccidial properties, are now being used in lower concentrations as growth promotants in food animals.

The name ionophore comes from the Greek, where *ion* is something that goes and *phore* means carrier. The ionophores are a group of naturally occurring monocarboxylic acid polyether antibiotics. Ionophores are able to form lipid soluble, reversible complexes with cations, and participate in transport and exchange of cations across biological membranes (Brandt, 1984). The result of this transport is an alteration in concentrations of cations and anions inside cells and the subcellular components that affect regulation of body functions. Oral administration of ionophores allows, initially, for regulation of body function at the site of digestion by positively affecting feed conversion efficiency and negatively affecting developmental stages of coccidia. As with any chemical introduced to food animals, it is important to understand the basis of veterinary use, the pharmacology (including the mechanism of action, absorption and metabolism), and single species and comparative toxicology. Resistance patterns, the residue profile and the depletion profile in edible tissues must also be understood.

[1]Director, Nonclinical Regulatory Affairs and Compliance, Research Triangle Park, North Carolina, USA.

8.2 PHARMACOLOGY

There are two major classes of ionophores. The *neutral ionophores* form charged complexes with cations to transport these complexes against the electrochemical gradient. The *carboxylic ionophores* form electrically neutral complexes with cations and promote an electrically neutral exchange diffusion of cations (Pressman and Fahim, 1982). The most widely used ionophores in veterinary medicine are monensin, lasalocid, narasin, and salinomycin. They are carboxylic ionophores produced from unidentified *Streptomyces* cultured from soil samples.

The unusual solubility properties of the ionophores may be explained by the oxygen functional groups that are concentrated in the center of the molecules, and the hydrophobic alkyl groups which are on the surface (Westley, 1977).

The two main functional properties that account for the mechanism of action of ionophores are the ability to interact and complex with metal ions, and the ability to carry ions across biological membranes. In the polar environment on the membrane surface, the anionic form of the ionophore is confined to the membrane interface where it is stabilized. As an anion, it has a negative charge, and is thus capable of binding with the positively-charged metal cations. This binding yields a lipid-soluble cyclic complex that can diffuse through the interior of the membrane. The complex, when reaching the opposite side of the membrane, is again subjected to a polar environment where the ionophore releases its enclosed cation and once again becomes acyclic until another cation complex is initiated. Depending upon the ionophore, the direction and degree of ion flux appears to be determined by the chemical gradients for the cations (e.g., Na^+, K^+, Ca^{++} and H^+). The rate of, and degree to which, a cation will be transported depends upon the affinity constant for the cation and the relative concentration gradient of the ion across the membrane (Bergen and Bates, 1984).

Monensin is monovalent and transports both sodium and potassium across cell walls, with a greater affinity for the sodium ion (Pressman and Fahim, 1982; Rumsey, 1984). Salinomycin and its methyl homologue, narasin, possess roughly equivalent preferences for Na^+ and K^+. Lasalocid is a divalent compound and can transport Ca^{++} and Mg^{++} as well as Na^+ and K^+ across cell walls. It has a preferential affinity for the potassium ion, with lesser but similar affinities for Na^+ and Ca^{++} (Pressman and Fahim, 1982; Rumsey, 1984). Of the neutral cations, lasalocid is one of the more disruptive to biological systems. It has a tendency to dimerize, and can form electrically neutral complexes with biologically important divalent ions such as Ca^{++} and Mg^{++} (Pressman and Fahim, 1982). Lasalocid has a wide range of complexation affinities and transport capabilities, encompassing not only inorganic polyvalent and alkali ions, but also primary amines such as catecholamines (Pressman and Fahim, 1982). Since it can transport both Na^+ and Ca^{++} effectively, it is difficult to determine which of these ions is primarily responsible for the many effects ascribed to this ionophore.

Pharmacological side effects, purportedly associated with the mechanism of action, include an increase in heart rate, cardiac output, and packed cell volume (Galitzer and Oehme, 1984).

8.3 PHARMACOKINETICS

In cattle, monensin absorption and metabolism have been studied extensively. When administered orally, it is absorbed from the gastrointestinal tract, rapidly distributed, metabolized (the exact metabolism is species dependent), excreted in the bile, and eliminated in the feces. At recommended doses, it is virtually undetectable in the blood, and it does not accumulate in most tissues of orally-dosed animals. When fed to cattle and chickens according to recommended practices, momensin was not detected (limit of detection 0.05 ppm) in edible tissues (Donoho, 1984).

Monensin has been shown to form more than 50 metabolites. The glucoside metabolite is active as a coccidiostat, as demonstrated by increased feed utilization in ruminants (Westley, 1977).

When ^{14}C-labelled monensin was given to steers, essentially all was eventually excreted in the feces after conversion to many metabolites that accumulate in the liver (Pressman and Fahim, 1982). In another study designed to investigate the metabolism of monensin using ^{14}C-labelled drug, none of the six metabolites isolated was responsible for more that 6% of the radioactivity (Donoho, 1978). Prolonged levels of parent lasalocid offer concrete evidence that lasalocid is absorbed.

Additional studies of pharmacokinetics and/or drug residues may be needed prior to approval where physiologic alterations in metabolism of an ionophore are suspected, or use of an ionophore in conjunction with another drug is expected. Knowledge that all ionophores are subject to extensive metabolism calls for expanded investigation of both disposition and residues under these circumstances.

8.4 TOXICITY

The neutral ionophores, such as monensin, lasalocid, salinomycin, and narasin — as charged molecules — are capable of grossly distorting biological membrane and action potentials. This presumably accounts for their high toxicity (Pressman and Fahim, 1982). The differences in the pharmacodynamic activity of the ionophores relating to monovalent and divalent cation complexing may also be related to the differences in toxicity (Comben, 1984). Additionally, there is considerable species variation in toxicity (Table 8.1).

Consistent signs of acute toxicity in animals were anorexia, hypoactivity, skeletal muscle weakness, ataxia, diarrhea, decreased weight gain and delayed death (Folkerts, 1982; Hanrahan et al., 1981).

With accidental overdosage of lasalocid or monensin, almost all cattle show time-dependent onset of symptoms and dose-dependent severity of symptoms. These include flank muscle tremors, increased heart rate, anorexia and diarrhea in all animals, followed by depression. Rapid breathing and ataxia may be seen only in severe overdose situations. Adverse findings observed on necropsy include pericardia and epicardial hemorrhage, pulmonary edema, and gastrointestinal reddening (Folkerts, 1982; Galitzer and Oehme, 1984). High concentrations of salinomycin and monensin cause cardiac arrhythmias in vivo via suppression of automaticity and transmembrane action potentials in isolated canine left ventricular Purkinje fibers, suppressed automaticity, and shortened action potential duration (Gaide et al., 1984). These measured responses are probably related to altered effects on cation transport.

Monensin has shown no evidence of carcinogenicity in lifetime rat studies. The primary target organs for chronic toxicity were skeletal and cardiac muscles. The cause of the lesions is not completely understood,

TABLE 8.1. Species Comparison of LD_{50} (mg/kg).

Species	Monensin	Reference	Lasalocid	Reference
Cattle	20–30	1, 2, 3	>35–<50	5, 6, 7
Chickens	200	2, 3	71.5	4
Dogs	>15*			2
Goats	26			2, 3
Horses	2–3	1, 2, 3	21.5	4
Monkeys	>160			2
Mice	78*, 125	2, 3	146	4
Rabbits	42	2	40	4
Rats	34*	2	122	4
Sheep	12	2, 3		
Swine	17	2, 3		
Trout	>1000	2		
Turkeys	200	2		

*Estimated average results reported for males and females.
Sources: Pressman and Fahim (1982), Todd (1984), Folkerst (1982), Galitzer and Oehme (1984), Brandy (1984), Galitzer et al. (1986), Galitzer et al. (1982).

TABLE 8.2. Approved Uses of Four Ionophores.

	Monensin	Lasalocid	Salinomycic	Narasin
CATTLE (not pastured)				
coccidiosis		360 mg/d*		
growth promotion	5–30 g/t 360 mg/d	10–30 g/t** 360 mg/d		
SHEEP				
coccidiosis		not approved		
growth promotion		20–30 g/t 70 mg/d		
BROILERS				
coccidiosis	90–110 g/t	68–113 g/t	40–60 g/t	54–72 g/t
TURKEYS				
coccidiosis	54–90 g/t		fatally toxic	fatally toxic

*milligrams/ton.
**grams/ton.
Source: 21CFR 500, 1988.

but it is also considered to be related to the altered transport of cations. Those muscles with the highest level of transport activity showed the most toxicity (Todd et al., 1984).

Workers involved in monensin production or monensin feed compounding have reported adverse reactions, such as headaches, nausea, nosebleeds, and skin rashes (Pressman and Fahim, 1982).

8.5 VETERINARY MEDICAL USE (UTILIZATION)

Ionophores are used, alone and in combination, as coccidiostats in cattle and poultry, and to promote growth and feed efficiency in cattle (Table 8.2). For treatment of coccidiosis, the ionophores affect the developmental stages of *Eimeria tenella* (coccidia, sporozoa), both in vivo and in vitro (Melhorn et al., 1983). The trophosporozoite, and perhaps the first generation schizont, are the stages attacked. The drug forms ionophoric complexes with sodium and potassium in the host and in the developing parasite. The ionophore/cation complex renders membranes permeable to cations, causing inhibition of certain mitochondrial functions, including substrate oxidations and ATP hydrolysis. The destruction of free merozoites occurs by bursting the pellicle (cell border), endoplasmic reticulum, and internal organelles after a very short exposure time of 20 minutes at 1, 10, and 100 ppm.

Because sporozoites, intracellular stages, and parasites are not as severely and rapidly affected, ionophores must be fed daily. Equipotency against 40,000 *Eimeria tenella* oocysts was shown at 70 ppm salinomycin, 125 ppm monensin, and 125 ppm lasalocid (Chappel and Babcock, 1979). It is not clear why parasites are selectively inhibited, but it may be related to the lack of active transport of sodium in the parasite.

In poultry, these drugs are often used in combination with roxarsone, virginiamycin, and/or the bambermycins to enhance their effectiveness. In calves, protective coccidiostatic effects were demonstrated by a decreased level of oocysts in calves fed medicated rations as low as 1.2 mg/kg of body weight (50 ppm) (Conlogue et al., 1984).

For improving weight gain and feed efficiency, the biologic responses related to the effectiveness of these ionophores include modification of volatile fatty acid production, modification of feed intake, change in ruminal gas production, changes in digestibility, changes in protein utilization, modification of rumen fill, rate of passage through the rumen, and some other proposed minor mechanisms (Schelling, 1984).

Normally, when carbohydrates are consumed, the forage cellulose is broken down into glucose by cellulose enzymes. If grain starch is consumed, amylase enzymes break down starch to glucose, which is further converted to pyruvate and then volatile fatty acids. The major portions of the volatile fatty acids are acetic acid, propionic acid and butyric acid, and the relative proportion of these acids depends upon the type of diet. Hay diets produce 3:1 acetic to propionic acid, and grain diets produce 1:1 acetic to propionic acid. Priopionate is more efficiently utilized by the ruminant because glucose breaking down into two 3-carbon propionate can enter the TCA cycle as succinyl CoA without losing any carbon units. Both acetate and butyrate must lose carbon in the breakdown as CO_2 (Olentine, 1982).

The carboxylic acid ionophores alter the rumen microflora by depressing microbial synthesis at concentrations between 11 and 66 ppm, to favor propionate production and decrease acetic acid concentrations. They have a greater effect where the diet is hay-based rather than grain-based. Monensin and lasalocid decrease gas and rumen methane production, while having no effect on the total volatile fatty acid concentration. This desirable effect may occur via the inhibition of deaminase, leading to the improved utilization of high-protein feed by changing the site of degradation from the rumen to the abomasum and the small intestine (chemical rumen bypass).

High grain rations produce a buildup of lactic acid, which can lead to acidosis. Ionophores, in addition to their coccidiostatic activity, inhibit the production of rumen bacteria that produce lactate. Thus, the occurrence of feedlot acidosis is controlled with the use of ionophores.

The specific exchange reactions catalyzed by the different ionophores in use with ruminants today depends on the cation affinity of the ionophore, the relative ion gradients, and the specific mechanism by which ion translocation occurs. Regardless of the exact nature of the event, interference with normal ion flux plays a dominant role in the mode of action of ionophores within the rumen (Bergen and Bates, 1984).

8.6 RESISTANCE

Chicken coccidia do not readily develop resistance to the polyether anticoccidial drugs where previous exposure had not occurred (Jeffers, 1981). Differences in sensitivity of broiler isolates to various ionophores have been reported. In one report, coccidia responded best to salinomycin and lasalocid where they had not previously been used on the farms. Monesin controlled coccidia better if the dose was increased (Mathis et al., 1984). In another study, narasin- and monensin-resistant strains were generally more sensitive to lasalocid. However, one strain was resistant to all three (lasalocid, narasin, and monensin) (Weppelman et al., 1977). Chapman (1986) reported that from *E. tenella,* a strain normally sensitive to monensin, chicken isolates have been identified that are resistant to all four commonly used ionophores. These reports indicate that in broilers coccidia appear to develop resistance to ionophores as a result of long-term exposure, and that coccidia that have never been exposed to ionophores retain their sensitivity (Mathis, 1984).

A laboratory strain of turkey coccidia, *E. meleagridis,* developed resistance to monensin within four generations. This is contrary to previous assumptions based on the difficulty that has been shown in producing resistance to ionophores with chicken coccidial strains (Jeffers and Bentley, 1980; Ryley, 1980). Rapid development of resistance has also been reported after treating *E. bovis* in calves, using monensin at a dose of 1 mg/kg. The same dose was not effective against a second infection produced 35 days after the first (Stockdale et al., 1983). When the sensitivity of 32 field isolates of *E. acervulina* to 100 ppm monensin was studied, results showed that with increased use of monensin, more resistant strains may eventually emerge (Chapman, 1982).

Use of ionophores at low levels have been associated with the development of resistant coccidial strains previously sensitive to ionophores. These strains are opportunistic, and prolific when sensitive strains are inhibited. With use of ionophores at low levels, development of resistance was identified in field-collected isolates, and subsequently produced in the laboratory where oocyst production was studied in the presence of drug. The basis for the resistance is presumably a basic change in the nature of the cytochrome b enzyme of the respiratory pathway in a sensitive strain. Cross-resistance was also demonstrated between lasalocid and monensin (Ryley, 1980).

8.7 RESIDUES IN EDIBLE ANIMAL TISSUES

The literature contains references describing the excretion and tissue distribution of [^{14}C] monensin in cattle and chickens (Herberg et al., 1978; Donoho et al., 1982). Tissue accumulation of ionophores is minimal, with no detectable residues in edible tissues other than liver reported at five days after the final dose (Herberg et al., 1978; Donoho et al., 1982). In chickens, at 6 hours after the last dose (zero-withdrawal), the liver contained 0.518 ppm monensin, and at 5 days after withdrawal, the average liver concentration was 0.106 ppm. In cattle, the liver residues averaged 0.4 ppm at 12 hours after the last dose (zero-withdrawal). The highest residue levels are found in the liver and the levels detected in other tissues are negligible in comparison to those of liver.

A tolerance (with respect to safe levels for human consumption) refers to the concentration of marker residues (e.g., a metabolite) in the target tissue. The marker residue is used to monitor for total drug residues in the edible tissues. A safe concentration refers to the total residue concentration in an edible tissue which is considered safe for human consumption. The tolerances, or safe concentrations, along with tissue residue depletion data, are used to determine an appropriate withdrawal period for a given drug at a recommended dose for each species. When used alone and not in combination, zero-withdrawal periods appear to be indicated for the ionophores, as they are not reported under the limitations set in 21CFR 556 1988 (Table 8.3). However, it is important to take into account that addition of the carboxylic ionophores to diets may change the bioavailability, gut uptake, and tissue deposition of divalent minerals (Elsasser, 1984).

8.8 ANALYSES AND ASSAYS

It is essential that assays be developed and validated for identifying residues of drugs used in food animals. Some published methods for the detection of residues of the ionophores are included in the Summary.

Liquid chromatographic techniques using fluorometric detection have been developed and validated for quantitation of beef liver tissue residues of monensin, salinomycin, lasalocid, and narasin (Martinez and Shimoda, 1985, 1986). Residues of the ionophores themselves have also been separated using thin-layer chromatography and bioautography. The literature

TABLE 8.3. Safe Concentrations for Total Residues of Ionophores Determined in Uncooked Edible Tissue in PPM.

		Monensin	Lasalocid	Salinomycin	Narasin
Chicken	muscle	1.5	1.2	NR	0.6
	skin + fat	3.0	2.4	NR	1.2
	liver	4.5	7.2	NR	1.8
Cattle	muscle	0.05*	1.2		
	fat	0.05*	4.8		
	liver	0.05*	4.8		
	kidney	0.05*	3.6		
Sheep	muscle		1.2		
	fat		6.0		
	liver		6.0		
	kidney		6.0		

* = only a tolerance is reported.
NR = not reported in the CFR.
Source: 21CFR 556, 1988.

describing assays for monensin includes an enzyme-linked immunosorbent assay (ELISA), an antibody assay for monensin, chemical assays, liquid chromatography/gas chromatographic systems, and TLC with bioautoradiography (Pressman and Fahim, 1982; Goras and Lacourse, 1984; Analytical Methods Comm., 1986; Weiss and MacDonald, 1985). The latter two assays are sensitive enough to quantitate tissue residue in chickens. For salinomycin, an HPLC determination has been reported which is sensitive enough for detecting and quantitating tissue residues, and a liquid chromatographic method for detecting and quantitating sodium salinomycin (which is not sensitive enough) has been reported.

8.9 SUMMARY

Ionophores have a unique niche in veterinary medicine. With ionophores in the diet, food animals are less prone to coccidiosis, they gain weight faster, and they consume less feed. They are also used in control of bloat and are toxic to fly larva in feces of ionophore-fed cattle (Goodrich et al., 1984). Proposed mechanisms for ionophoric activity have been related to effects on cellular cation transport, either in the animal or in the parasite. At high doses, the toxicities observed are apparently logical extensions of the pharmacological mechanism of action. With low-level use, selective resistance to ionophoric coccidiostatic activity has developed, along with cross-resistance.

Ionophores are used primarily in food-producing animals. In some species, metabolites of ionophores have been shown to accumulate in the liver after oral administration. These metabolites are used as marker residues to determine when tissue residues of ionophores are below the tolerance, and to set withdrawal periods. Where ionophores are poorly absorbed, no withdrawal period may be authorized.

8.10 REFERENCES

Analytical Methods Committee. 1986. "Collaborative Studies of Methods for the Detection of Residues on Monensin in Chicken Tissues," *Analyst*, 111(9):1089–93.

Bergen, W. G. and D. B. Bates. 1984. "Ionophores: Their Effect on Production Efficiency and Mode of Action," *J. of Animal Sci.*, 58(6):1465–68.

Brandt, W. 1984. "Veterinary Aspects of Ionophores in Feedlot Cattle, in the Use of Drugs in Food Animal Medicine," in *Proceedings of 10th Annual Food Animal Medicine Conference*, J. D. Powers and T. E. Powers, eds., Ohio State University Press, pp. 281–300.

Chapman, H. D. 1986. "Isolates of *Eimeria tenella:* Studies on Resistance to Ionophorous Anticoccidial Drugs," *Res. Vet. Scie.*, 41(2):281–282.

Chapman, H. D. 1982. "The Sensitivity of Field Isolates of Eimeria Acervulina Type Monensin," *Vet. Parasitol.*, 9(3-4):179–83.

Chappel, L. R. and W. E. Babcock. 1979. "Field Trials Comparing Salinomycin (Coxistac), Monensin, and Lasalocid in the Control of Coccidiosis in Broilers," *Poult. Sci.*, 58(2):304–7.

Comben, N. 1984. "Toxicity of the Ionophers (Letter)," *Vet. Rec.*, 114(5):128.

Conlogue, G., W. J. Foreyt and R. B. Wescott. 1984. "Bovine Coccidiosis: Protective Effects of Low-Level Infection and Coccidiostat Treatments in Calves," *Am. J. Vet. Res.*, 45(5):863–866.

Donoho, A. L. 1984. Biochemical Studies on the Fate of Monensin in Animals and in the Environment," *J. of Animal Sci.*, 58(6):1528–1539.

Donoho, A. L., R. J. Herberg, L. L. Zornes and R. L. Van-Duyn. 1982. "Excretion and Tissue Distribution of [^{14}C] Monesin in Chickens," *Agri. Food. Chem.*, 30(5):909–913.

Donoho, A. L., J. Manthey, J. Occolowitz and L. Zornes. 1978. "Metabolism of Monensin in the Steer and Rat," *Agri. and Food Chem.*, 26(5):1090–1095.

Elsasser, T. H. 1984. "Potential Interactions of Ionophore Drugs with Divalent Cations and Their Function in the Animal Body," *J. Anim. Sci.*, 59(3):845–853.

Folkerts, T. M. 1982. "Monensin," in *Current Veterinary Food Animal Practice*, J. C. Howard, ed., W. B. Saunders Co., p. 400.

Gaide, M. S., J. K. Gibson and A. L. Bassett. 1984. "Cellular Basis for Arrhythmogenicity of Ionophores with Different Cation Selectives," *Eur. J. Pharmacol.*, 99(4):349–352.

Galitzer, S. J., E. E. Bartley and F. W. Oehme. 1982. "Preliminary Studies on Lasalocid Toxicosis in Cattle," *Vet. Hum. Toxicol*, 24(6):406–409.

Galitzer, S. J. and F. W. Oehme. 1984. "A Literature Review on the Toxicity on Lasalocid, a Polyether Antibiotic," *Vet. Hum. Toxicol*, 26(4):322–326.

Goodrich, R. D., J. E. Garrett, D. R. Gast, M. A. Kirick, D. A. Larson and J. C. Meiske. 1984. "Influence of Monensin on the Performance of Cattle," *J. of Anim. Sci.*, 58(6):1484.

Goras, J. T. and W. R. Lacourse. 1984. "Liquid Chromatographic Determination of Sodium Salinomycin in Feeds, with Post-Column Reaction," *J. Assoc. Off. Anal. Chem.*, 67(4):701–706.

Halvorson, D. A., C. Van-Dijk, and P. Brown. 1982. "Ionophore Toxicity in Turkey Breeders," *Avian Dis.*, 26(3):634–639.

Hanrahan, L. A., D. E. Corrier and S. A. Nagi. 1981. "Monensin Toxicosis in Broiler Chickens," *Vet. Pathol.*, 18(5):665–671.

Herberg, R., J. Manthey, L. Richardson, C. Cooley and A. Donoho. 1978. "Excretion and Tissue Distribution of [^{14}C] Monensin in Cattle," *J. Agric. Food Chem.*, 26(5):1087–1090.

Jeffers, T. K. and E. J. Bentley. 1980. "Experimental Development of Nonensin Resistance in *Eimeria Meleagrimitis*," *Poult. Sci.*, 59(8):1731–1735.

Jeffers, T. K. 1981. Resistance and Cross-Resistance Studies with Narasin, a New Polyether Antibiotic Anticoccidial Drug," *Avian Dis.*, 25(2):395–403.

Martinez, E. E. and W. Shimoda. 1985. "Determination of Monensin Sodium Residues in Beef Liver Tissue by Liquid Chromatography of a Fluorescent Derative," *J. Assoc. Off. Anal. Chem.*, 68(6):1149–1153.

Martinez, E. E. and W. Shimoda. 1986. "Liquid Chromatographic Determination of Multiresidue Fluorescent Derivatives of Ionophore Compounds, Monensin, Salinomycin, Narasin, and Lasalocid, in Beef Liver Tissue," *J. Assoc. off. Anal. Chem.*, 69(4):637–641.

Mathis, G. F., L. R. McDougald and B. McMurray. 1984. "Drug Sensitivity of Coccidia from Broiler Breeder Pullets and from Broilers in the Same Integrated Company," *Avian Dis.*, 28(2):453–459.

Mehlhorn, H., H. Pooch and W. Raether. 1983. "The Action of Polyether Ionophorous Antibiotics (Monensin, Salinomycin, Lasalocid) on Development States of *Eimeria tenella* (Coccidia, Sporozoa) in vivo and in vitro: Study by Light and Electron Microscopy," (abst) - *Z. Parasitenkd.*, 69(4):457–471.

Olentine, C. 1982. "Ionophores for Beef Cattle," *Feed Management* (December):14–19.

Pressman, B. B. and M. Fahim. 1982. "Pharmacology and Toxicology of the Monovalent Carboxylic Ionophores," *Ann. Rev. Pharmocol. Toxicol.*, 22:465–490.

Rumsey, T. S. 1984. "Monensin in Cattle: Introduction," *J. Anim. Sci.*, 58(6):1461–1464.

Ryley, J. F. 1980. "Drug Resistance in Coccidia," *Adv. Vet. Sci. Comp. Med.*, 24:99–120.

Schelling, G. T. 1984. "Monensin Mode of Action in the Rumen," *J. of Animal Sci.*, 58(6):1518–1520.

Stockdale, P. H., A. Sheard and G. B. Tiffin. 1982. "Resistance to *Eimeria bovis* Produced after Chemotherapy of Experimental Infections in Calves," *Vet. Parasitol.*, 9(3-4):171–177.

Todd, G. C., M. N. Novilla and L. C. Howard. 1984. "Comparative Toxicology of Monensin Sodium in Laboratory Animals," *J. of Animal Sci.*, 58(6):1512–1517.

Weiss, G. and A. MacDonald. 1985. "Methods for Determination of Iophonore-Type Antibiotic Residues in Animal Tissues," *J. Assoc. Off. Anal. Chem.*, 68(5):971–980.

Weiss, G., N. R. Felicito and M. Kaykaty et al. 1983. "Tissue Residue Regulatory Method for the Determination of Lasalocid Sodium in Cattle Liver Using High Performance Liquid Chromatography with Fluorometric Detection," *J. Agric. Food Chem.*, 31(1):75–78.

Weppelman, R. M., G. Olson, D. A. Smith, T. Tamas and A. Van-Iderstine. 1977. "Comparison of Antococcidial Efficacy, Resistance and Tolerance of Narasin, Monensin and Lasalocid in Chicken Battery Trials," *Poult., Sci.*, 56(5):1550–1559.

Westly, J. W. 1977. "Polyether Antibiotics: Versatile Carboxylid Acid Ionophores Produced by Streptomyces," *Applied Microbiology*, D. Perlman, ed., Academic Press, pp. 177–223.

CHAPTER 9

Parasiticides

T. B. BARRAGRY, Ph.D.[1]

9.1 INTRODUCTION

MODERN research has provided us with a considerable array of highly effective and highly selective drugs with which to attack helminth parasites. It is of paramount importance that these agents be used correctly and judiciously if a favorable clinical response to treatment is to be obtained.

Most of the older anthelmintic compounds had rather narrow margins of safety. Modern drugs possess considerably improved activity against the immature or larval stages of parasites, have a higher margin of safety, and a broader spectrum of activity. Many factors limit the usefulness of anthelmintics in practice. These may be features relating to the inherent efficacy of the drug itself, its mechanism of action, its pharmacokinetics properties, features relating to the host animal (such as the operation of the oesophageal groove reflex), or features relating to the parasite (such as its location in the body, its degree of hypobiosis or whether it has acquired anthelmintic resistance).

Before discussing general principles of anthelmintic activity and the individual compounds, attention should be drawn to some of the major characteristics to look for in a new anthelmintic agent.

(1) It should have a broad spectrum of activity against mature, immature and, if possible, arrested larvae.

(2) It must be easy to administer to a large number of animals.

[1]Department of Veterinary Surgery and Companion Animal Medicine, Faculty of Veterinary Medicine, University College, Ballsbridge, Dublin, Ireland.

(3) It must have a wide margin of safety, and be compatible with other compounds.

(4) It should not present a residue problem requiring long withdrawal periods.

(5) The compound should be economical for the livestock producer to use.

An anthelmintic must be selectively toxic to the parasite. This is usually achieved by either the inherent pharmacokinetic properties of the compound itself (causing the parasite to be exposed to higher concentrations of the anthelmintic than the host cells), or by inhibition of metabolic processes vital to the parasite.

9.1.1 MECHANISM OF ACTION OF ANTHELMINTIC

Although the primary physiological mode of action of all drugs used against parasites is not fully understood, it is nonetheless possible to indicate general sites of action and biochemical mechanisms.

Life support mechanisms of parasites are based mainly on maintaining an advantageous feeding site, and using the acquired foodstuff to generate chemical energy. Parasitic helminths must maintain an appropriate feeding site, and nematodes and trematodes must actively ingest and move food through their digestive tracts to maintain an appropriate energy state.

These functions require proper neuromuscular coordination. The major common pharmacological basis of the treatment of helminths generally involves interference with one or more of these functions vital to the parasite, i.e., disruption of energy process and subsequent starvation of the parasite, or neuromuscular coordination, paralysis of the parasite and subsequent expulsion by peristalsis (see Table 9.1).

TABLE 9.1. Benzimidazoles and Related Compounds.

Generic name	R^1	R^2	Molecular Formula
Albendazole	$CH_3CH_2CH_2-S-$	$-NH-CO_2CH_3$	$C_{12}H_{15}N_3O_2S$
Cambendazole	$(CH_3)_2CH-O-CO-NH-$	4-Thiazolyl	$C_{13}H_{14}N_4O_2S$
Fenbendazole	C_6H_5-S-	$-NH-CO_2CH_3$	$C_{15}H_{13}N_3O_2S$
Flubendazole	$p-F-C_6H_4-CO-$	$-NH-CO_2CH_3$	$C_{16}H_{12}FN_3O_3$
Mebendazole	C_6H_5-CO-	$-NH-CO_2CH_3$	$C_{16}H_{13}N_3O_3$
Oxfendazole	C_6H_5-SO-	$-NH-CO_2CH_3$	$C_{15}H_{13}N_3O_3S$
Oxibendazole	$CH_3CH_2CH_2-O-$	$-NH-CO_2CH_3$	$C_{12}H_{15}N_3O_3$
Parbendazole	$CH_3CH_2CH_2CH_3$	$-NH-CO_2CH_3$	$C_{13}H_{17}N_3O_2$
Thiabendazole	H	4-hiazolyl	$C_{10}H_7N_3S$

9.1.2 ENERGY PROCESSES

Several chemical classes of compounds and many different types of metabolic inhibitors are to be found within the group of commonly used anthelmintics in veterinary practice.

Included are inhibitors of mitochondrial reactions (fumarate reductase) and/or glucose transport, the benzimidazoles, which include: albendazole, cambendazole, parbendazole, mebendazole, fenbendazole, oxibendazole, oxfendazole, thiabendazole, and flubendazole. The pro-benzimidazole thiophanate and febantel are metabolized in vivo to active benzimidazole and thus act in the same manner.

Another class of compounds is comprised of uncouplers of oxidative phosphorylation, including salicylanilides (oxyclozanide, rafoxanide, niclosamide, clioxanide, closantel brotianide), and substituted phenols (nitroxynil, bithional, niclofolan, hexachlorophene).

The benzimidazoles inhibit the enzyme fumarate reductase, and thus inhibit energy generation. These compounds are broad spectrum in activity, and are often effective against adult larvae and eggs. Being chemically quite similar, and affecting the same metabolic pathway in the parasite, cross resistance frequently exists amongst the benzimidazoles.

Mebendazole and flubendazole interact with the substance tubulin in the intestinal cells of nematodes, resulting in the disappearance of microtubule from the cells, and decreased absorption and digestion of nutrients such as glucose.

In the case of all these benzimidazoles it has been shown that one very important factor in ensuring their efficacy is the prolongation of contact time between the drug and the parasite. The requirement that this contact be prolonged for as long as possible is a consequence of the mode of action of this group. Since they act upon the energy systems within the nematode cell, death occurs after the stored energy sources are exhausted. This is a slow process compared to the much quicker actions of other anthelmintics which act on the neuromuscular coordination system of the parasite.

Uncoupling of oxidative phosphorylation processes has been demonstrated for a number of compounds, especially the fasciolicides. Many of these compounds are ineffective in vivo against nematode species of parasites. This ineffectiveness appears to be due to a permeability barrier, since isolated nematode mitochondria are susceptible. Because these compounds are general uncouplers of oxidative phosphorylation, their safety indices are not as high as the benzimidazole, although they are adequate if used as directed. Looseness of feces and slight loss of appetite may be seen in some animals after treatment at the recommended dose rates. High dosages may cause blindness, hyperthermia, convulsions, and death—classical symptoms of uncoupled phosphorylation.

9.1.3 NEUROMUSCULAR COORDINATION

Interference with neuromuscular coordination in the parasite may occur by inhibiting the breakdown of excitatory neurotransmitters, or by mimicking the action of the excitatory transmitter, resulting in spastic paralysis of the parasite. Other mechanisms mimic the action of the inhibitory transmitter, or cause hyperpolarization, with an ensuing flaccid paralysis of the parasite. Either spastic or flaccid paralysis of an intestinal helminth allows the normal peristaltic action of the host to expel the parasite.

Cholinesterase inhibitors are organophosphates which cause spastic paralysis and include haloxan, trichlorfon dichlorvos, coumaphos, etc.

Cholinergic agonists include imidothiazoles (tetramisole, levamisole) and pyrimidines (morantel, pyrantel). Piperazine produces muscle hyperpolarization.

Avermectins work through the potentiation of inhibitory transmitters. Thus, anthelmintic action is closely related to the unique life-support parasite requirements. Drug efficacy is generally based upon interference with energy generation of neuromuscular coordination. Differential toxicity between host and parasite is based on the presence of a unique parasite system, or on an effective concentration of drug that will inhibit the parasite system without interfering with the comparable host system (selective toxicity).

9.1.4 ADMINISTRATION OF ANTHELMINTICS

Anthelmintics can be administered in a variety of ways. In general, drench, paste and injectable preparations allow a greater degree of control over the amount of anthelmintic that is administered to an individual animal than in feed or medicated block preparations. Whichever method of administration is used, it is important to read the manufacturer's instructions with particular regard to:

- the types of worm against which the anthelmintic is active
- the class of stock for which the product is recommended, and any limitations to its use which may be advised
- the dose rate to be used, and any increase in dose rate that may be recommended to deal with different development stages, or different species or types of worm
- the withholding period

The solubility of an individual anthelmintic compound largely governs the choice of route of administration. Insoluble anthelmintics usually have to be given orally in the form of suspensions, pastes or granules. The more soluble compounds may be given orally as solutions, topically in the pour-on form (organophosphates, levamisole), or subcutaneously as an injectable solution (nitroxynil, rafoxanide, levamisole). Particle size is an important consideration in determining the behavior and subsequent efficacy or toxicity or an orally-administered agent. In general, small particle size will determine the rate and extent of absorption from the gastrointestinal tract and may increase the efficacy of compounds that must undergo metabolism in the liver before exerting their anthelmintic effect. Conversely, large particle size, coupled with insolubility, will minimize gut absorption, may reduce systemic toxicity and will ensure increased efficacy of the parent drug against rumen-dwelling parasites further down the alimentary tract.

The route of administration of these drugs also determines their persistence in the body and accordingly their antiparasitic efficacy. In ruminants, administration directly into the abomasum by way of the oesophageal groove bypass may increase the rate of absorption and excretion of an anthelmintic. This resultant lack of persistence may reduce the efficacy of the compound by reducing the period of contact between the drug and the parasite. Lowered efficacy may occasionally occur due to an unexplained direct passage of the drug into the abomasum, and a more rapid flow of drug through the alimentary tract. Operation of the rumen bypass acts to reduce the efficacy of certain benzimidazole anthelmintics, for instance oxfendazole. It has been deduced experimentally that immediate arrival of the compound in the abomasum following dosing reduced its efficacy from 91% to 45% against a thiabendazole-resistant strain of *H. contortus.*

The absorption of levamisole is not affected by the route of administration since it is highly soluble, is unaffected by the operation of the ruminal bypass and thus is superior to benzimidazole in this particular respect. In general, most of the benzimidazoles are most effective if injected into the abomasum. Deposition in the rumen tends to prolong the passage of the rather insoluble compound through the alimentary tract, slows the rate of absorption, and so maintains the level of circulating anthelmintic. The rumen, in fact, acts as a type of drug reservoir, from which plasma concentrations can be sustained for long periods, and which also slows the passage of unabsorbed drug through the alimentary tract.

As with injectable preparations, high blood levels of the anthelmintic are achieved rapidly and they offer the advantage of ease of administration.

9.1.5 IN-FEED PREPARATION

Many of the benzimidazoles are available for mass medication through in-feed preparations. This method allows little control over the amount of anthelmintic that individual animals consume, unless they are fed separately or in small supervised groups. There is therefore a risk that some animals will receive less than the recommended dose for their weight and that others will receive more. However, on account of the high safety margin of the benzimidazole this should not be a problem.

9.1.6 DOSING PASTE SYRINGES

Many anthelmintics for equine use are presented in the form of paraffin-based oral dosing syringes. The precalibrated amount of paste is deposited carefully on the animal's tongue. This is now a very popular method in horses, as it is less troublesome than stomachtubing, and is more convenient and labor saving.

9.1.7 SUSTAINED RELEASE FORMULATIONS

Slow release devices represent a new approach to the way in which anthelmintic may be administered, and the manner in which they can be used to control parasitic disease.

9.1.8 BOLUSES

With all parasiticide boluses, retention in the rumen is critical for efficacy. To minimize the risk of boluses being coughed up and lost, a critical density is incorporated into the bolus design to ensure the weighted delivery device remains continually in the gut.

The objective of controlled-release delivery systems is to modify the release rate of a drug so that the normal exponential plasma decay curve is not attained. These new delivery systems are characterized by rate-controlled drug release profiles capable of producing relatively constant concentrations in the bloodstream or other sites in the body. This predetermined release rate can be rendered constant for periods of hours, days or months, exposing the recipient's tissue to constant drug levels for extended exposure periods. The peak and valley plasma fluctuations associated with conventional dosage forms in concentration-time profiles are thus avoided.

Considerable latitude exists with respect to the design of oral sustained-delivery systems for administration to ruminant species, where the unique anatomical features of the alimentary tract facilitate the entrapment of controlled release devices. The critical concern in the design of any suitable ruminant bolus system is to ensure retention of the drug reservoir within the reticulorumen and to prevent its elimination by the process of regurgitation or intestinal expulsion. Ruminal retention is achieved either on the basis of density, or the variable geometrical configuration of the device. The minimum size is determined by the dimension of the oesophageal and omasal orifices. Although sustained-release rumen devices for mineral and trace element supplementation have been used for a number of years, it is only relatively recently that pharmaceutical technology has developed methods of ensuring reliable delivery of organic drugs, such as anthelmintic, from such systems. With conventional anthelmintic treatment, repeat dosing is necessary, which offers no preventative or residual effect, and so animals may be reinfested relatively soon after dosing. Controlled release of anthelmintic from sustained-delivery rumen systems provides a useful alternative approach whereby continuous anthelmintic output protects animals from overwintered pasture infestation, and prevents a subsequent pasture buildup of eggs and infective larvae. For effective prophylactic or therapeutic applications, a reliable, well-designed delivery system must contain a potent agent which is effective at relatively low dosage levels.

The morantel sustained-release bolus is a cylindrical metal bolus comprising an inner drug reservoir and a microporous barrier wall permeable to drug passage by diffusion. The pores of the material in contact with the drug reservoir contain a hydrogel, and provide controlled and predictable release rates of drug for an extended period of time. The bolus is loaded with 13.5 g of morantel tartrate in polyethylene glycol. The active principle escapes across the cellulose acetate end-discs at a constant rate of approximately 90 mg/day for the duration of the grazing season. The bolus is given prophylactically at turnout, and the continuous low level release of morantel suppresses worm egg output by treated animals, reduces larval infection on pasture and prevents appearance of clinical parasitism. Strategic use of the bolus disrupts the normal epidemiological pattern of bovine gastrointestinal nematodiasis.

Oxfendazole is available in the form of a pulsed release bolus for use in cattle between 100 kg and 200 kg body weight. The bolus is designed to release oxfendazole not continuously but intermittently as a full therapeutic dose (750 mg) five times at approximately 21-day intervals. The bolus device is made up of a PVC segment, five individual cells, a corroding central alloy core and a steel endweight to prevent regurgitation from the rumen. Each individual cell comprises a circular tablet containing 750 mg oxfendazole and a silicone rubber sealing washer. In the rumen, corrosion of the central steel-magnesium alloy spine leaves the PVC caps between the oxfendazole tablets unsupported. As each divider drops off, an oxfendazole tablet is left exposed. This process is repeated at approximately 23-day intervals, giving a life span of 130 days to the bolus. It is claimed that pulsed release systems possess fewer problems than the continued release bolus, where constant subtherapeutic levels of anthelmintic are leached out.

The bolus is a compact polymeric bolus containing 22 g of levamisole. Encased in an outer impermeable shell, the polymeric matrix becomes slowly impregnated with water, increasing the internal surface area and ultimately expelling the levamisole through the single outer pore.

A sustained release formulation of ivermectin is currently being developed for use in ruminant animals. To date, osmotic pump delivery systems containing ivermectin have been investigated with a view to ensuring rate-controlled release of ivermectin over prolonged periods of time.

9.2 BENZIMIDAZOLES

Benzimidazole parasiticides comprise the largest chemical family of veterinary drugs used for the treatment of endoparasitic diseases in domestic animals. They are characterized by broad spectrum activity against a range of parasites and high safety indices in target species. Their high degree of clinical efficacy under field conditions is related both to their pharmacodynamic and pharmacokinetic characteristics. The use of benzimidazole anthelmintics gives some cause for concern from the standpoint of drug residue formation in animal fluids and tissues. This is on account of their slow kinetic excretory pathways, their complex biotransformation, the varied type of metabolites generated in the target species, the potential biological significance of the residues and metabolites, and the clearly demonstrated toxicological properties of a number of the metabolites and parent compounds. The fact that parasitic disease in food animals is an increasing worldwide problem, and that benzimidazole anthelmintics are routinely used by the layman (often at times of peak production) compounds the potential residue hazard for the human consumer.

The benzimidazole group is derived from the simple benzimidazole nucleus, and includes the thiabendazole analogs and the benzimidazole carbamates. Substitution of various sidechains and radicals on the parent nucleus gives rise to the individual members of this group. Generally substituents on position R^2 of the nucleus tend to influence the inherent efficacy of the compound while substituents on R^1 affect primarily the pharmacokinetic disposition profile. Modification of the kinetics of a particular benzimidazole will indirectly affect clinical potency, because such factors as relative insolubility, slow elimination and persistent circulation of parent drug and/or active metabolites will increase the drug-parasite contact time. Because of the pharmacodynamic mechanism of action of these drugs, such prolongation of contact time increases both antiparasitic efficacy and spectrum of use.

Thiabendazole, the first commercially available benzimidazole, was introduced in 1961 for use in domestic animals. Metabolism of this prototype compound occurs by way of hydroxylation to the 5-hydroxy compound, which is then rapidly eliminated in urine as the glucuronide. Efforts to prolong the sojourn of this substance in the body (to amplify potency) led to the introduction of a substituent at the 5 position to prevent rapid hydroxylation. From this development, the related compound cambendazole was produced. Further studies on the chemical classes of substituents paved the way for newer benzimidazole molecules characterized by longer half-lives in the animal body, distinct biotransformation pathways and greater potency (Table 9.1). Inherent in this tendency towards slower body clearance is the necessity to observe longer tissue and milk withholding times.

Most of the modern benzimidazoles possess a carbamate substituent on position R^5. Variations in efficacy between these carbamates is principally a function of their differing kinetics.

A number of benzimidazoles exist in the form of prodrugs. These owe their anthelmintic activity to the fact that they are metabolized in the animal body to the biologically active benzimidazole carbamate nucleus. Febantel, netobimin and thiophanate are examples of such prodrugs. Synthesis of benzimidazole carbamates depends on a cyclization process. Metabolic or chemical activation generates an active benzimidazole in vivo from an inactive prodrug precursor. Cyclization occurs in the case of thiophanate. Febantel is hydrolysed to the active metabolite fenbendazole. Netobimin undergoes a process both of reduction and cyclisation to yield the active compound albendazole (McDougall et al., 1985; Steel et al., 1985; Delatour et al., 1986).

Netobimin is a nitrophenylguanidine which is cyclized to albendazole following reduction of the nitro group to an amino group (Steel et al., 1985; Delatour et al., 1986) and febantel is a phenylguanidine which is hydrolyzed by removal of a methoxyacetyl group and then cyclized to fenbendazole (Delatour et al., 1982).

The differences in kinetics of the benzimidazoles, especially the sulphoxide metabolites that are thought to be the most anthelmintically active forms, between sheep, cattle and goats account for the different efficacy of the drugs in each species, and this the higher dosages recommended for cattle (Anon., 1983) and goats (Bogan et al., 1987a) for single administration. Benzimidazoles display very large kinetic differences between ruminant and monogastric species.

Anatomical features such as the rumen and caecum of ruminant and equine species, respectively, which slow the passage of digesta in these species, may enhance the activity of the benzimidazoles by extending their transit time. This has been confirmed in sheep following oral and intraabomasal administration of fenbendazole, whereby the drug maintains high plasma concentrations for a prolonged period when given orally (Marriner and Bogan, 1981b).

For most members of the group, extensive biotransformation occurs primarily in the liver, although extra-hepatic sites may be involved in metabolic conversion. Metabolism takes the form of oxidative or hydrolytic cleavage processes, resulting in free water-soluble metabolites. These metabolites concentrate in the plasma, tissues, bile and urine. Biliary secretion of metabolites and conjugates may occur, and the more polar the metabolites the more readily they are excreted in the urine. Elimination through the milk and feces is observed also; some enterohepatic circulation takes place (Hennessy, 1985).

Biological activity can be associated with the parent drug alone, its metabolites, or in some cases both. For thiabendazole and cambendazole, anthelmintic activity resides with the parent drug nucleus alone.

In the case of some of the benzimidazole carbamates, the processes of biotransformation and metabolic interconversion are more complex. Metabolites are associated with anthelmintic activity and biological reactivity. This is especially true of many of the sulfoxide and sulfone derivatives. For prodrugs such as febantel, netobimin and thiophanate, biological activity resides in the resultant benzimidazole carbamate metabolites produced by metabolic conversion in the body.

More recently developed benzimidazole carbamates are characterized by novel substituents on the R^5 position of the benzimidazole nucleus, and the replacement of the thiazole ring by methylcarbamate. Such molecular modifications have spawned a new benzimidazole generation, with much slower rates of elimination, higher potencies, broader spectra and complex metabolic pathways. Inherent in these developments is the attendant risk of greater biological reactivity of their metabolites, and the possibility of their heightened toxicological significance.

For the most potent benzimidazoles, i.e., albendazole, fenbendazole, and oxfendazole, it would appear that activity is related to the sulfoxide metabolite. Fenbendazole owes much of its activity to the sulfoxide metabolite (oxfendazole) formed from it. The principal route of metabolism for albendazole, fenbendazole and oxfendazole appears to be oxidation of the sulfide to the sulfoxide and sulfone metabolites (Bogan et al., 1982, 1983, 1984). Activity is associated with both the sulfide and sulfoxide which are metabolically interconvertible. In vitro incubation of sulfide (fenbendazole) or sulfoxide (oxfendazole) with bovine hepatic microsomes readily yields the sulfoxide or sulfide respectively, i.e., fenbendazole and oxfendazole are interconvertible. Ruminal fluid can cause reduction of the sulfoxide to the sulfide. Febantel, fenbendazole, and oxfendazole follow similar metabolic patterns in the body (Marriner and Bogan, 1981a; Prichard, 1978; Delatour, 1982, 1985). Febantel is the prodrug of both these benzimidazoles. (Fenbendazole is a sulfide and oxfendazole is the corresponding sulfoxide.) Albendazole is oxidized to the sulfoxide and both albendazole and its sulfoxide possess similar anthelmintic and embryotoxic properties. The sulfoxides of albendazole, fenbendazole and oxfendazole are further oxidized to the sulfones. The pharmacokinetics of FBZ in serum or plasma have been examined in a variety of mammalian species following oral or intragastric administration. Several of these studies have also identified and/or quantified the four known metabolites of FBZ. The major metabolite is fenbendazole sulfoxide (FBZ-SO; oxfendazole) which also possesses anthelmintic activity. Other metabolites include fenbendazole sulfone (FBZ-SO$_2$) p-hydroxyfenbendazole (FBZ-OH) and the product of demethoxycarbonylation, fenbendazole amine (FBZ-NH$_2$). Considerable information is available on the disposition and elimination of FBZ and its metabolites. Very little of the parent drug or any of its metabolites is excreted in the urine of any species. Small quantities of FBZ-NH$_2$ have been reported in urine of cattle and pigs (Duwel, 1977). Likewise, the sulfoxide (FBZ-SO) and sulfone (FBZ-SO$_2$) metabolites are present in feces and urine in very small quantities, bringing into question the importance of FBZ-SO (oxfendazole) as an active anthelmintic metabolite in the lower gastrointestinal tract.

While it is generally assumed that the oxidative metabolism of FBZ occurs in liver, there are no reports which have documented the rate or pathways of metabolism in vitro. Marriner and Bogan (1981a) cited unpublished observations that the metabolization of oxfendazole (FBZ-SO) was carried out by hepatic microsomal enzymes. All of the metabolites of FBZ except FBZ-NH$_2$ are produced in the liver. The hydrolysis of FBZ to FBZ-NH$_2$ is a very minor pathway, and while this metabolite had been demonstrated to occur in plasma of goats (Short et al., 1987a) and, presumably, in urine of cattle and pigs (Duwel, 1977), its formation does not seem to result from the action of a hepatic esterase.

The rates of metabolite production probably do not reflect maximal rates of metabolism for each pathway because aromatic hydroxylation and sulfoxidation are competing activities and because oxidation of the sulfoxide (FBZ-SO) to the sulfone (FBZ-SO$_2$) is a two-step series-linked reaction.

Fenbendazole is readily metabolized to its sulfoxide, oxfendazole (FBZ-SO), which is also an active anthelmintic (Averkin et al., 1975). This oxidation occurs in the liver (Marriner and Bogan, 1981b) and may also occur in other organs or tissues of mammals as well. The sulfoxide can be further oxidized to the sulfone, FBZ-SO$_2$, a less active metabolite (Averkin et al., 1975).

Two other metabolites are known. One is the product of p-hydroxylation (FBZ-OH). The other is a metabolite resulting from demethoxycarbonylation referred to as fenbendazole amine (FBZ-NH$_2$). The activity of these metabolites has not been established, but the latter has been shown to be of very minor importance (Short et al., 1987a, 1987b).

It is well known that FBZ-SO can be reduced to FBZ, and both compounds appear in plasma shortly after the iv or oral administration of either one (Ngomuo et al., 1984; Prichard et al., 1985; Short et al., 1987a,

1987b). Marriner and Bogan (1987a) have shown that this reduction occurs in the rumen of sheep.

In all species studied, the principal route of metabolism appears to be by oxidation of the sulphide to the sulphoxide and the sulphone metabolites. Activity is associated with both the sulphide and sulphoxide, but appears to be minimal or absent in the sulphone. However, the sulphide-sulphoxide metabolism can be shown to be reversible. Ruminal fluid causes reduction of sulphoxide to sulphide only. After in vitro incubation of sulphide (fenbendazole) or sulphoxide (oxfendazole) with bovine hepatic microsomes, the presence of sulphoxide of sulphide, respectively, can be readily demonstrated. The rate constants for this conversion are difficult to determine, because the exceedingly low solubilities of fenbendazole and oxfendazole make it difficult to prepare accurate concentrations in aqueous solution when fenbendazole or oxfendazole is administered to sheep, the ratio of sulphoxide to sulphide in plasma samples is about 4:1.

Abomasal fluids were generally lower than were those of oxfendazole (limit of detection in plasma was 0.01 mg/mL). Concentrations of oxfendazole in the plasma of calves after oxfendazole administration have been reported (Nerenberg, Runken and Martin, 1978). However, the age and weight of these calves were not given, The mean peak concentrations of fenbendazole and its sulphoxide, oxfendazole, and sulfone metabolites were studied in sheep after oral administration (10 mg/kg). The mean peak concentrations found in plasma of fenbendazole, oxfendazole and sulfone were 0.15, 0.29 and 0.17 mg/mL at 24, 30 and 36 hours after administration, respectively. Fenbendazole and oxfendazole were detectable in plasma for 5 days after administration. Most of the anthelmintic activity of fenbendazole is due to the oxfendazole metabolites. Oxfendazole, in a test of anthelmintic efficacy in mice, was more active than were fenbendazole and the sulfone (Averkin et al., 1975). In such a test, it was difficult to measure the true anthelmintic activity of fenbendazole, because oxfendazole formed from it exerts anthelmintic activity. In the sheep, concentrations of oxfendazole were generally higher than were those of fenbendazole. Also, because the sulfone of fenbendazole was less active anthelmintically than was fenbendazole or oxfendazole, and because concentrations were relatively low, anthelmintic activity in sheep is greatly influenced by this metabolite. Thus, in the plasma and abomasal fluid of sheep, much of the anthelmintic activity must be due to oxfendazole formed from fenbendazole. As with albendazole, the sulfoxide (oxfendazole) is the result of hepatic metabolism, and its presence in the abomasum is due to passive diffusion from the systemic compartment. Concentrations of oxfendazole in plasma after fenbendazole administration is about 40% of those found after the same dose of oxfendazole in the same six sheep.

Benzimidazoles act by inhibiting the uptake of low-molecular-weight nutrients by the parasite which then subsequently starves (Tables 9.2a and 9.2b). This is brought about by a binding process between the active benzimidazole and tubulin, a structural protein (Hoebeke et al., 1976; Ireland et al., 1979; Friednam and Platzer, 1978). Blockage of polymerization of tubulin into microtubules damages the integrity and function of the absorptive cells within the parasite (Vanden Bossche and Nollin, 1973). Enzymes of energy metabolism, such as fumarate reductase, are also inhibited by benzimidazoles which prevent the parasite from absorbing and utilizing its foodstuffs. The antiparasitic effect is accordingly a lethal, but relatively slow, process.

Binding of benzimidazoles to tubulin is a reversible and saturable effect. An association has been established between antitubulin activity and teratogenicity for a number of compounds, including colchicine, griseofulvin, and vincristine. Mutagenic activity can also be associated with antitubulin activity. Antimitotic and genotoxic effects have been documented for a number of benzimidazoles (Lapras et al., 1975: Delatour et al., 1975; Mochida et al., 1983; Havercroft et al., 1981; Seiler, 1973; Kappas et al., 1976; Styles and Garner, 1974). In light of experimental work revealing genotoxic effects, such as chromosome doubling, the question of potential carcinogenicity must be addressed.

The high safety margin of benzimidazoles for target animals is due to the greater selective affinity for parasitic tubulin than for mammalian tissues. Nonetheless, some toxic effects based on antimitotic activity (teratogenicity/genotoxicity) do occur in target species. Hence, the selective toxicity of benzimidazoles is not absolute.

Parbendazole is the primary benzimidazole for which teratological effects (skeletal malformations) have been demonstrated in sheep (Schreiner and Holden, 1983; Lapras et al., 1973; Szabo et al., 1974). Albendazole, cambendazole, febantel and oxfendazole are also teratogenic in this species (Delatour, 1975, 1977; Johns, 1977; Clement, 1982) whereas fenbendazole, mebendazole and oxibendazole are not (Becker, 1975; Vanden Bossche, 1982; Theodorides, 1977). Teratogenic effects in target species occur at relatively lower dosage rates than those associated with acute toxicity. Species differences occur in domestic animals and in laboratory animals, as regards susceptibility to teratogenic effects. Dosage rate, specific benzimidazole, and stage of embryonic development are major influencing factors. Cattle seem to be unaffected by the benzimidazoles which are teratogenic in sheep (Wetzel, 1985). A similar relationship exists between rats and rabbits. Many benzimidazoles teratogenic in sheep are also teratogenic in rats, and the types of embryotoxic effects are similar.

TABLE 9.2a. General Spectrum of Activity of Some Commonly Used Anthelmintics in Ruminants.

Group	Chemical Name	G.I.T. Nematodes	Inhibit Larvae	Lungworm	Fluke
Benzimidazoles	Thiabendazole	+	-	-	-
	Parbendazole	+	+ -	-	-
	Fenbendazole	+	+	+	-
	Oxfendazole	+	+	+	-
	Albendazole	+	+	+	+
	Triclabendazole	-	-	-	+
Pro Benzimidazoles	Febantel	+	+	+	-
	Thiophanate	+	+ -	-	-
	Netobimin				-
Imidothiazoles	Levamisole	+	-	+	-
Pyrimidines	Morantel (Bolus)	+	Prevents Establishment	Prevents Establishment	-
Avermectins	Ivermectin	+	+	+	-
Ogano Phosphates	Haloxan	+	-	-	-
Piperazine	Diethylcarbamazine	-	-	+	-
Salicylanilides	Rafoxanide	-	-	-	+
	Oxyclozanide	-	-	-	+
	Closantel	-	-	-	+
Substituted Phenols	Nitroxynil	-	-	-	+
	Niclofolan	-	-	-	+
	Bithionol Sulphoxide	-	-	-	+
	Hexachlorophene	-	-	-	+
Aromatic Amide	Diamphenethide	-	-	-	+
	Clorsulon	-	-	-	+

TABLE 9.2b. Chemical Groups and Mechanisms of Action of Commonly Used Anthelmintic Agents.

Group	Chemical Name	Mode of Action in Parasite	Effect
Benzimidazoles	Thiabendazole Parabendazole Cambendazole Mebendazole Oxibendazole Fendendazole Albendazole Oxfenbendazole	Interfere with Energy Production Inhibit Fumarate Reductase Block Tubulin Synthesis Inhibit Glucose Transport	Starvation of Parasite (slow process) Ovicidal
Pro-Benzimidazole	Thiophanate Febantel Netobimin	Metabolized in vitro to Benzimidazole Carbamates	As Above
Imidothiazoles	Tetramisole Levamisole	Ganglionic Stimulants	Spastic Paralysis
Pyrimidines	Pyrantel Morantel	Cholinergic Agonists (Ganglionic)	Spastic Paralysis
Salicylanilides	Rafoxanide Oxyclozanide Niclosamide	Uncoupled Oxidative Phosphorylation	Energy Depletion
Substituted Phenols	Nitroxynil Niclofolan Bithionol Hexachlorophene	Uncoupled Oxidative Phosphorylation	Energy Depletion
Organo-Phosphate	Trichlorphon Dichlorovos Haloxan	Cholinesterase Inhibition	Spastic Paralysis
Piperazine	Piperazine Diethylcarbamizine	Neuromuscular Hyperpolarizers	Flaccid Paralysis
Avermectin	Ivermectin	GABA Potentiation	Flaccid Paralysis

139

Although benzimidazoles all possess similar pharmacodynamic effects (antitubulin), species differences in terms of sensitivity to embryotoxic effects are attributable to metabolic and pharmacokinetic disposition factors. Benzimidazoles can be classified as:

- those which are not teratogenic per se and have no teratogenic metabolite (oxibendazole)
- those which are not teratogenic and give rise to a teratogenic metabolite (fenbendazole)
- those which seem to be teratogenic per se and have no teratogenic metabolite (oxfendazole)
- those which are possibly teratogenic per se and have a metabolite which seems more toxic than the parent compound (mebendazole)

Thiabendazole is rapidly absorbed and metabolized by the liver. In cattle and sheep, peak plasma levels are attained in approximately 4 hours with metabolites accounting for up to 30% of the total concentration. The major primary metabolite is the 5-hydroxy component, which further conjugates with sulfate and glucuronide to increase water solubility and the rate of renal clearance (Tocco, 1965). The duration of pharmacological activity of thiabendazole is quite short. Nonetheless, using [^{35}S] thiabendazole, levels are still detectable in the brain, heart, liver, kidney and pancreas of sheep 5 days after dosing. At 16 days post-dosing, radioactivity is detectable in the liver and muscle (Tocco, 1964). Similar findings have been reported for calves and pigs. Using [^{35}S] thiabendazole in calves, radioactivity associated levels have been detected in the liver 59 days after administration (Tocco, 1965). This hepatic incorporation is indicative of metabolic incorporation into the endogenous pool, rather than drug storage, binding and residue burden deposition.

The metabolism and excretion of thiabendazole is more extensive in cattle than in sheep. The bovine species seems to possess greater capacity for oxidative metabolism for benzimidazole parasiticides (Marriner and Bogan, 1980). Systemic anthelmintic activity is greater in sheep than in cattle for thiabendazole (Weir, 1985). Clinically, this is exemplified by the fact that although thiabendazole is ineffective against lungworm infestation in cattle, it does possess useful activity against the ovine lung parasite. The brief duration of action of thiabendazole in animals is attributable to its solubility and rapid metabolism to pharmacologically inactive metabolites (Ngomuo, 1984).

There are very few reports of tissue and/or milk residue levels of benzimidazoles and their metabolites after their administration to domestic animals. Residues of benzimidazoles are greatest in the liver, and are also detected for the longest in this tissue. Thus, liver tissue should receive paramount attention in residue surveys.

Oxfendazole, in a test of anthelmintic efficacy in mice, was more active than were fenbendazole and the sulfone (Averkin et al., 1975). In such a test, it was difficult to measure the true anthelmintic activity of fenbendazole, because oxfendazole formed from it exerts anthelmintic activity. In the sheep, concentrations of oxfendazole were generally higher than were those of fenbendazole. Also, because the sulfone of fenbendazole was less active anthelmintically than was fenbendazole or oxfendazole, and because concentrations were relatively low, anthelmintic activity in sheep is greatly influenced by this metabolite. Thus, in the plasma and abomasal fluid of sheep, much of the anthelmintic activity must be due to oxfendazole formed from fenbendazole. As with albendazole, the sulfoxide, oxfendazole, is the result of hepatic metabolism, and its presence in the abomasum in plasma after fenbendazole administration is about 40% of those found after the same dose of oxfendazole in the same six sheep.

Oxfendazole levels in the milk of cows following administration of 5 mg/kg oxfendazole were not detected beyond 72 hours using HPLC (Tsina and Matin, 1981). Levels at 48 hours were 0.173 mg/mL; levels at 72 hours were 0.009 mg/mL.

Fenbendazole administered to cattle at a dosage rate of 10 mg/kg gave residue levels of 0.03 mg/mL in milk 1–2 days following administration. In liver, levels declined from 7.9 ppm (after 2 days) to 0.1 ppm at 14 days (Watson, 1983).

This is in line with the work of Duwel (1977) who examined fenbendazole tissue residues in cattle. At a dosage rate of 10 mg/kg, this worker noted maximum residues in the liver, the principal site of biotransformation. Fourteen days after oral treatment (FBZ 10 mg/kg) the liver level had declined to 0.1 mg/kg (1000 mg/kg). At 2 days post-dosing, fat and muscle tissue residues were 0.84 mg/kg and 0.34 mg/kg respectively, i.e., 9–23 times less than liver levels. By day 14 these declined to trace amounts.

Residues of albendazole, febantel, fenbendazole, oxfendazole, and thiabendazole concentrate primarily in liver tissue in cattle and sheep, and persist longest in this tissue. At therapeutic dosage rates, residue levels in liver exceed those of muscle and kidney. In sheep, oxfendazole reaches peak plasma concentration at 30 hours, while albendazole sulfoxide peaks at 20 hours. Residues are more persistent for febantel, fenbendazole, and oxfendazole than for albendazole or thiabendazole. Although residues are detectable in milk for these drugs, their presence is relatively shortlived. For example, the use of oxfendazole in cattle is associated with low levels of residues in milk. Using high pressure liquid chromatography (HPLC) after oral administration at 5 mg/kg oxfendazole is not detectable beyond 72 hours (Tsina, 1981).

Following dosing of cattle with 2.5% suspension of febantel at a dosage rate of 7.5 mg/kg, febantel sulfone, fenbendazole and oxfendazole were

detectable in milk. At 12 hour levels were 0.16 ppm for the parent compound, 0.035 for fenbendazole and 0.105 ppm for oxfendazole. At 60 hour
post-dosing, levels were approximately 0.01 ppm (Delatour, 1983).

Nine metabolites of albendazole were identified in the urine of cattle and
sheep which were given the labelled drug orally. Metabolic conversions
included oxidation at sulfur, alkyl and aromatic hydroxylation, methyla-
tion at both nitrogen and sulfur and carbamate hydrolysis. All detectable
urinary metabolites were sulfoxides or sulfones; unchanged albendazole
was present in only minor amounts (Gyurik, 1981).

In various tests, triclabendazole showed no mutagenic effects, and no
embryotoxicity or teratogenicity was observed in rats, sheep and cattle.
After oral administration triclabendazole is satisfactorily reabsorbed and
biotransformed, the two main metabolites formed being the sulphoxide
and sulphone derivatives. More than 95% of the quantity administered
orally is eliminated in the feces, about 2% in the urine and less than 1%
in the milk. After oral doses of 10 mg/kg in sheep and 12 mg/kg in cattle,
the parent and its metabolites are detectable in edible tissue (muscle, liver,
kidneys) over a prolonged period. The residue concentrations in edible
tissue drop to below 0.4 mg/kg within 28 days in sheep and 14 days in
cattle.

Parbendazole ^{14}C administered orally to sheep at 45 mg/kg resulted in
peak plasma levels of ^{14}C activity at 6 hours. Plasma levels dropped continuously until 48 hours at which time concentrations < 1 μg/mL were
detected. Seven urinary metabolites were identified. Liver residues
remained elevated at 1.41 ppm after 16 days (Dicuollo, 1974).

Poor absorption and low tissue concentrations were recorded in lambs
given [^{14}C] mebendazole. Liver concentration was highest, and radioactivity in this organ was still detectable at 15 days post-dosing (Burgat-
Sacaze et al., 1981). After oral administration, mebendazole remains
restricted to the gastrointestinal tract because of its poor absorption. A
very low amount is absorbed into the systemic circulation and distributed
to the tissues. In the intestinal tract, mebendazole is present for its largest
part as unchanged parent compound (more than 80%). In the gut itself,
metabolism of mebendazole does not occur. The largest amounts are found
in the liver with maximal radioactivity levels, representing 4% of the dose,
present at 2 to 4 hours after dosing. The radioactivity is mainly due to
metabolites, so that it can be stated that the liver is the main site for metabolism. This is in common with most other benzimidazoles.

Studies of the metabolism of [^{14}C] mebendazole after oral administration in lambs showed poor absorption and low tissue concentrations,
except in the liver (see Table 9.3). Tissue total residue concentrations fell
rapidly. At day 15 and after, radioactivity persisted only in the liver; expressed as mebendazole equivalent, it amounted to 1.07 ppm, i.e., 0.13%
of the dose (Burgat-Sacaze et al., 1981).

TABLE 9.3. Mebendazole Metabolism in Lambs (Rico, 1981).

Organs	Days after Dosing		
	2 (48 hours)	10	15
liver	24.5*	3.1	10.7
kidney	6.9	0.09	0.0
muscle	0.4	0.0	0.0

*Total radioactivity expressed as mebendazole equivalents (ppm).

In the case of [^{14}C] cambendazole, the radioactivity was detectable in calves' liver 30 days after administration, and a fraction of this was in the bound form (Baer, 1977).* Table 9.4 summarizes milk and tissue residue data from a number of experiments.

Cambendazole and mebendazole are specific examples of benzimidazoles which form bound residues. With cambendazole it is difficult to assess the precise nature of the bound form. The principal binding site is through interaction of the thiazole moiety with endogenous compounds. The toxicological assessment of this nonextractable fraction is difficult. Residual radioactivity is also dependent on the type and position of radiolabelling. Bound residues of parbendazole found in the liver of sheep after 15 days have been attributed to incorporation of radioactivity into amino acids. In the case of thiabendazole, liver radioactivity is also found in proteins, lipids, nucleic acids and glycogen. In recent years, an increasing number of drugs have been reported for which metabolism includes some covalent binding of metabolites to macromolecules. Several veterinary drugs belong to this list, and give rise to nonextractable bound residues in tissues. This tissue-bound residue is of considerable interest, because of the possible participation of this binding process in target species toxicity, as well as its potential toxicological significance in terms of food hygiene for humans.

The binding of reactive metabolites to biological macromolecules has been implicated in several toxic manifestations including carcinogenicity (Miller and Miller, 1976; Swenson et al., 1974), allergenic effects (Parker, 1982), hepatotoxicity (Jollow et al., 1973), nephrotoxicity (Mitchell et al., 1973), neutrotoxicity (de Caprio et al., 1982), lung toxicity (Boyd and Burka, 1978), bone marrow toxicity (Irons et al., 1980), and finally teratogenicity (Martz et al., 1977; Gordon et al., 1981). The mechanism of covalent binding of metabolites to macromolecules generally involves a preliminary metabolic oxidative reaction. However, it is not possible to

*Non-extractable (bound) residues—the fraction of radioactivity which persists in tissues for weeks or months.

TABLE 9.4. Tissue and Milk Residue Levels of Benzimidazoles.

Drug	Dose (mg/kg)	Time Post-dosing	Tissue	Level (ppm)	Reference
Albendazole	20 (Cattle)	24 hours	Liver	436	Delatour et al. (1981)
		4 days	Liver	191	
Febantel	7.5 (Cattle)	48 hours	Milk	0.06	Delatour et al. (1983)
		60 hours	Milk	0.03	Delatour et al. (1983)
Fenbendazole	10 (Cattle)	14 days	Liver	0.1	Watson (1983)
Thiabendazole	50 (Cattle)	3 days	Muscle	0.08	Watson (1983)
	50 (Sheep)	16 days	Liver	0.15	Watson (1983)

144

draw conclusions about the toxicological significance of bound residues from absolute values of residue levels.

Early studies on the mode of action of benzimidazole anthelmintics indicated that they disrupt normal glucose metabolism by inhibition of key enzymes such as fumerate reductase (Prichard, 1973). More recent studies, however, have demonstrated that a number of benzimidazoles have an ability to bind to tubulin, and thus interfere with normal microtubule function (Friedman and Platzer, 1987; Kohler and Bachmann, 1981; Watts, 1981; Ireland et al., 1982). Many benzimidazoles also prevent normal embryonation and hatching of nematode and trematode eggs, a process dependent in part on normal microtubule function (Kelly and Hall, 1979; Coles and Brisco, 1978). Fenbendazole, through its metabolite, oxfendazole, possesses embryonic potential.

Toxicological assessment of bound residues can be attempted by bioavailability studies and relay toxicity methods. With albendazole- and cambendazole-bound residue, bioavailability is low. Experiments in rats demonstrated that following administration of calf liver treated with cambendazole only 15% of total radioactivity was absorbed. Relay embryotoxicity studies for these two benzimidazoles in rats following feeding of liver residue tissue revealed no toxicity. Bioavailability of albendazole-bound residue tissue revealed no toxicity. Bioavailability of albendazole-bound residues is 4–5% from beef liver (Grannec, 1980; Scott and Di Cuollo, 1980).

The bioavailability and toxicity of nonextractable residues often differ from those of the parent compounds. Bound residues may contain degradation products of the parent compound in intermediary metabolism, or new compounds resulting from covalent binding of the parent drug or metabolites to endogenous macromolecules.

As is true of many xenobiotics, electrophilic reactants arise from biotransformation processes. This is also true of benzimidazoles, and resultant electrophilic metabolites undergo nonextractable binding. The basic benzimidazole ring structure is relatively stable. However, for thiabendazole and cambendazole, thiazole ring metabolism leads to incorporation into endogenous metabolism and hepatic tissue protein (Rosenblum et al., 1971; Baer et al., 1977). Binding at the hepatic site has also been noted with albendazole (Delatour et al., 1983).

Nonextractable residues, although toxicologically potentially significant, could also theoretically be considered relatively nontoxic for a number of reasons. Once bound, it is possible that the fixed residue may no longer possess biological reactivity and become "unbound" (Delatour and Parish, 1986). On the other hand, where the risk involves allergenicity, bound residues may possess the potential to trigger hypersensitivity reactions. Distinguishing between drug-related residues bound to macromolecules, and fractions which have entered the normal metabolic pool is difficult.

Extractable residues** of benzimidazoles undoubtedly constitute a relatively higher toxicological risk. The free metabolites account not only for the primary and desirable anthelmintic activity but also for the observed undesirable secondary toxic effects (embryotoxicity, teratogenicity), in the case of specific compounds. A biological relationship exists between many metabolites and teratogenicity. Albendazole is embryotoxic in rats when the dosage exceeds 6 mg/kg. The urinary sulfoxide metabolites display embryotoxic properties, and these have bioavailability of 74%. In contrast, the bioavailability of bound residues in edible tissue is 3–4.4% (Burgat-Sacaze et al., 1981).

Although extractable residues (parent drug or free metabolites) can be correlated with definitive toxicological potential, safety margins can nonetheless be defined. To determine the acceptable daily intake (ADI) in food, therefore, is an extremely complex procedure for benzimidazole residues. While the ADI is computed on the basis of the lowest no effect level (NEL) in the most sensitive species studied, toxicologists and regulatory agencies differ in their interpretation of the most suitable safety margins applicable. Some countries adopt a safety margin of up to 2,000 for drugs which are known or possible teratogens. Although ADI's can be determined, international agreement is necessary to standardize safety margins, and to harmonize toxicological criteria. Withdrawal periods for benzimidazoles are calculated on the basis of metabolic and toxicological data, taking into account the bioavailability of the residues.

The approach taken may be twofold. Either establish a withdrawal time so that residues are no longer detectable or establish that the existing detectable residues are not deleterious to human health. This involves calculation of the ADI, because the NEL's are low for benzimidazoles. Teratogenic metabolites have been identified and measured in edible animal products.

9.3 SALICYLANILIDES AND SUBSTITUTED PHENOLS

All the members of the three chemical groupings possess clinical efficacy against liver flukes.† The salicylanilides and substituted phenols act

---

** Extractable residues — the fractions which include all compounds and metabolites in free form or loosely bound in tissues.

† *Salicylanilides:* brotianide, closantel, oxyclozanide and rafoxanide. *Substituted phenols:* bithionol, disophenol, hexachlorophene, niclofolan and nitroxynil. *Aromatic amide:* diamphenethide.

to uncouple or disconnect the mitochondrial reactions involved in electron transport associated events from ATP generation (Rew, 1978). The adult flukes are affected mainly in vivo with variable activity against the immature flukes in the liver parenchyma. The lowered efficacy of a number of the salicylanilides and substituted phenols against the immature flukes may be due to the high protein binding of the drugs in the blood which bathes the immature stages in the liver tissues. A number of these compounds, however, do possess activity against 6 week old flukes in cattle and sheep. Metabolism may affect the pharmacological activity of various fasciolicides, and some of the metabolism may occur in the gastrointestinal tract. For example, nitroxynil is metabolized by rumen bacteria, which destroys its activity and restricts administration to injection.

In the case of diamphenethide, an aromatic amide which is given orally, metabolism may be important for full efficacy. Following absorption diamphenethide is further metabolized in the liver to an amine metabolite which is active against flukes. It is not active against liver flukes in vitro unless incubated in the presence of enzymatically functional liver cells. In contrast, oxyclozanide is metabolized in the liver to anthelmintically-active glucuronide and is excreted in the bile in high concentrations in the vicinity of the adult fluke. Biliary excretion is a very important pathway for fasciolicides active against the adult parasite in the bile ducts. Most of the available fasciolicides are administered in the form of oral suspensions, or occasionally as solutions by subcutaneous injection. Tables 9.5 and 9.6 list some available flukicides for ruminants.

Increasing the dosage of these compounds above the therapeutic rate frequently results in increased activity against the later parenchymal stages, but where such elevated dosages are employed, cognizance must be taken of the inherent safety margin of the particular drug in use. That the bile ducts are important in the excretion of many of these compounds is evidenced by the high proportion of these and their metabolites excreted in the feces rather than the urine. The increased susceptibility of the developing flukes is largely related to the pharmacokinetic behavior of the individual compounds. Since they are bound to plasma proteins, many of the individual compounds are slowly excreted, producing long plasma half-lives, which means lengthy withholding times must be observed in many cases (Lee, 1973). The fasciolicidal activity of salicylanilides is dependent upon the extent to which they persist in the plasma. For example, rafoxanide is fully absorbed, and plasma levels of approximately 29 mg/mL are found 24 hours after sheep receive the recommended dose. The plasma half-life is about 4 days and it binds to plasma proteins with a very high affinity. Plasma will dissolve up to 2 mg/mL, whereas it is virtually insoluble in water. Relatively high levels of residues are found in plasma even 42 days after dosing, but residues in other tissues are negligible. The plasma

TABLE 9.5. Anthelmintics for Use against *Fascioloa hepatica* in Sheep.

Chemical Group	Drug	Efficacy > 90%			Withholding Times (days)
		1–4 Weeks	4–8 Weeks	Adult	Meat
Aromatic Amide	Diamphenethide	+	+	+/−	7
Benzimidazoles	Albendazoles	−	−	+	14
	Triclabendazole	+	+	+	28
Salicylanilides	Oxyclozanide	−	−	+	14
	Rafoxanide	−	+/−	+	28
Substituted phenols	Niclofolan	−	−	+	14
	Nitroxynil	−	+/−	+	30

TABLE 9.6. Anthelmintics for Use against *Fascioloa hepatica* in Cattle.

Chemical Group	Drug	Efficacy > 90%			Withholding Times (days)	
		6 Week Flukes	6-10 Week Flukes	Adults	Meat	Milk
Benzimidazoles	Albendazole	−	−	+	14	3
	Triclabendazole	+	+	+	14	0
Salicylanilides	Oxyclozanide	−	−	+	14	0
	Rafoxanide	−	+	+	21/28	0
Substituted phenols	Niclofolan	−	−	+	14	0
	Nitroxynil	+ / −	+	+	30	3

half-life of disophenol is even longer, being 7–14 days in dogs, and in excess of 30 days in sheep. The penalty for high efficacy associated with long plasma half-life is the need for long withholding periods.

The high efficacy of many salicylanilides and substituted phenols against blood ingesting parasites, such as *H. contortus* and hookworms, may be related to their attachment to plasma proteins. Presumably they are released to poison the parasite after ingestion of blood. Pharmacokinetic data for many modern fasciolicides is sparse.

Peak plasma levels which may be an indicator of efficacy are reached in 12–24 hours for salicylanilides. The absorption of substituted phenols, such as nitroxynil given parenterally, is rapid and complete, peak plasma levels being achieved 30–60 minutes after dosing. The relatively high residues of nitroxynil found in milk are due to the relatively high dosage, parenteral administration and the tendency to form stable complexes with serum and body proteins (Table 9.7). Nitroxynil is retained in the liver and plasma of sheep at detectable levels (> 0.1 ppm) for 66 days after a single dose of 10 mg/kg. Although binding to serum albumin occurs, long term exposure to the drug is critical for antiparasitic activity.

Clorsulon is a sulfonamide which inhibits certain glycolytic pathways of trematodes. Given orally at the rate of 4 mg/kg clorsulon is 91–96% effective in the treatment of 56-day old *Fasciola hepatica* infections. In addition to effective fluke removal clorsulon also displays a stunting-effect in the injectable formulation for liver fluke infection in cattle. Immature flukes (8 weeks old) require a higher dose (7 mg/kg) than adult flukes (3.6 mg/kg). In plasma, clorsulon is bound to protein which, when ingested by liver flukes, inhibits enzymes of the glycolytic pathway. The salicylanilides (closantel, oxyclozanide and rafoxanide) have long biological half-lives in sheep (14.5, 6.4 and 16.6 days respectively). These long half-lives are related to the very high plasma protein binding of these three drugs (> 99%). Residues in liver are detectable for extended periods following administration (Mohammed-Ali and Bogan, 1987).

Both salicylanilides and substituted phenols are potent uncouplers of oxidative phosphorylation. Their selective toxicity for the parasite is determined by their rate of absorption and metabolism, the pathways of excretion, their affinity for plasma protein and their persistence in the bloodstream. Because they are general uncouplers of oxidative phosphorylation, their safety indices are usually not as high as many other anthelmintic agents, but nonetheless are more than adequate if used as directed. Commonly, a slight loss of appetite and looseness of feces may be seen following treatment at the recommended dosage rates. High dosage rates may cause blindness and classical symptoms of uncoupled phosphorylation, i.e., hyperventilation, hyperthermia, convulsions, tachycardia and ultimately death. Adverse effects are most commonly seen

TABLE 9.7. Milk Residue Levels of Fasciolocides.

	Niclofolan	Oxyclozanide	Nitroxynil
Peak conc. excreted (mg/kg)	0.06	0.13	1.46
Time until analytical detection is reached (hrs.)	168	168	216
Ratio of total dose to conc. in milk at 120 hours	600:1	500:1	56:1
Ratio of milk conc. at 120 hours to 0.01 mg/kg	1:1	2:1	20:1

A residue of 0.01 mg/kg and less is taken to be toxicologically insignificant.

Source: W. Heeschen, 1979.

in severely stressed animals, in animals in poor condition nutritionally or metabolically, or in animals with very severe parasitic infestations.

9.4 IMIDOTHIAZOLES—LEVAMISOLE

Levamisole is the laevo isomer of di-tetramisole, which is a racemic mixture. The parent compound tetramisole was first marketed as an anthelmintic in 1965 but it was soon noted that its anthelmintic activity resided almost entirely in the L-isomer, levamisole. Thus it was determined that the dosage could be reduced by half using the L-isomer alone. Reducing the dosage in this way also appreciably increased the margin of safety, while leaving the anthelmintic potency unchanged. Levamisole is widely used as an anthelmintic agent, and although it is less toxic than tetramisole, nonetheless some preparations of tetramisole are still available. Levamisole possesses a broad spectrum of activity against a wide range of gastrointestinal helminths and lungworms. It is commonly used in cattle, sheep, pigs, goats and poultry and very occasionally in the horse, dog and cat. It is normally administered orally or by subcutaneous injection in the soluble hydrochloride form, and in general the two routes are considered equivalent in efficacy. Topical preparations for cattle have recently been introduced onto the market. Levamisole slow-release boluses are available in some countries.

Levamisole is a ganglion stimulant of nematode nerves, leading to neuromuscular paralysis of the parasites. (Hexamethonium, a ganglionic blocker, inhibits the action of levamisole). Because it acts on the roundworm nervous system, it is not ovicidal. Its broad spectrum of activity against nematodes, ease of use, water solubility, reasonable safety margin and lack of teratogenic effects have allowed it to be used successfully in a wide range of hosts. Peak blood levels of levamisole occur within 30 minutes of administration. These concentrations then decline over a period of 6–8 hours with 90% of the total dosage being excreted in 24 hours, largely in the urine. After oral dosing in cattle, levamisole residues are present in muscle, fat, liver and kidney at 2 hours post-medication but are below the limits of assay detectability (0.1 ppm) at 48 hours after treatment. Blood and urine levels of levamisole peak at 2–6 hours post-medication but are below the 0.1 ppm level of detectability at 36 and 72 hours after treatment respectively (Simkins, 1975). Muscle, fat, liver, kidney, blood and urine contain levamisole at 2 hours post-medication following injectable levamisole. No residues are detectable 7–8 days after injection in the tissues, blood or urine. Highest residue levels appear in the liver.

Levamisole hydrochloride residues in milk averaged 0.50, 0.55, 0.58 and

0.32 ppm 12 hours after the administration of levamisole drench (8 mg/kg), feed bolus, and injectable formulation. Residues were below 0.01 ppm in milk at 48 hours after drench treatment, and at 60 hours after treatment with the other three formulations (Simkins et al., 1972).

In sheep, levamisole at a dose rate of 7.5 mg/kg produced mean plasma concentrations of 3.1, 0.7 and 0.8 μg/mL after administration by the subcutaneous, oral and intra-ruminal routes respectively (Bogan et al., 1982).

Levamisole residues in edible tissues from sows given 8 mg/kg orally, were < 0.1 mg/kg of muscle and fat in sows killed in post-treatment day 3, and < 0.1 mg/kg of kidney in sows killed on post-treatment day 5. Liver residues averaged 0.78 mg/kg in sows killed on day 3, and were reduced to 0.31 mg/kg in sows killed on day 5 (Berger et al., 1987).

Thus comparing the routes of administration, it would be expected that subcutaneous, rather than oral, administration would provide better efficacy against lungworms by virtue of the higher plasma (and therefore lung) concentrations achieved. In addition, because of its different mechanism of action, levamisole possesses activity against benzimidazole-resistant parasites.

In contrast to the benzimidazoles, excretion is rapid and the consequent withholding times are therefore shorter than with the insoluble benzimidazoles. Mammalian toxicity with levamisole, especially with tetramisole, is usually greater than for the benzimidazoles, although toxic signs are not usually seen unless the normal therapeutic dosage is exceeded. Levamisole toxicity in the host animal is largely an extension of its antiparasitic effect, i.e., cholinergic-type symptoms of salivation, muscle tremors, ataxia, urination, defecation and collapse. In fatal levamisole poisoning, the immediate cause of death is asphyxia from respiratory failure. Atropine sulfate can alleviate such symptoms in the host animal.

Levamisole may cause some inflammation at the site of subcutaneous injection, but this is usually of a transient nature. The compound is widely used in cattle, sheep, pigs and poultry for all the major parasites. It possesses good activity against adult and immature parasites of the GIT and lungs. It has no activity against fluke and tapeworms. Its margin of safety is sufficiently narrow in the horse, cat and dog to preclude use in these species, except under special conditions.

Levamisole has been found experimentally to possess immunostimulant effects, and it has recently been used clinically in humans and, to a limited extent, in animals in a variety of disease conditions. The use of levamisole in the treatment of brucellosis in man has been described, and experimental work has shown that levamisole exerts its immunostimulant effects by restoring the number of T-lymphocytes to normal when these are depleted. Several clinically measurable parameters of immune status are restored to normal by the drug. The significance of the immunological potential of levamisole residues has yet to be assessed.

On account of its mechanism of action, the peak concentration of levamisole achieved in the blood, rather than the duration of concentrations, is more relevant to its antiparasitic activity.

9.4.1 LEVAMISOLE STUDIES ON IMMUNOLOGICAL EFFECTS

There are several reports regarding the effects of levamisole on established tumors or tumor cells in experimental animals. Levamisole failed to activate macrophages against syngeneic tumor cells in Balb/c mice after intraperitoneal injection. However, mammary tumors induced by dimethyldibenzanthracene in rats showed regression with levamisole when this was given orally. The effect was dose-dependent and it was seen where immunopotentiation was also observed. The beneficial effects, where noted, have been ascribed to the potentiation of cellular immunity by levamisole.

9.4.2 OBSERVATION IN HUMANS

Various nonspecific effects have been noted after humans have been given therapeutic doses of levamisole. These have included nausea, vomiting, abdominal pain, taste disturbances, fatigue, headache, confusion, dizziness, fever, insomnia, arthralgia, muscle pain, hypotension, vasculitis and skin rashes.

Administration of levamisole has resulted in both neutropenia and thrombocytopenia.

One of the most common, and probably the most severe, side effects of levamisole in humans is agranulocytosis. The highest incidence was noted in patients with rheumatoid arthritis.

There are no data on which to base a dose response or NEL for levamisole-induced agranulocytosis in humans. One study indicated that even with low doses of levamisole, agranulocytosis could develop if the duration of dosing was prolonged.

Thus, levamisol-induced agranulocytosis appears to be an idiosyncratic reaction which, in some instances, may be associated with HLA-B27 seropositive rheumatoid arthritis or otherwise abnormal immune function. It is reversible, although corticosteroids may be required to initiate the recovery process.

9.5 ORGANOPHOSPHATES

A number of organophosphates have come to be used as anthelmintics in recent years. While they were originally extensively employed as insecticides, nowadays agents such as dichlorvos, trichlorfon and haloxan are commonly used as anthelmintics in sheep, cattle, horses and dogs. Dichlorvos and trichlorfon are particularly effective against bots and ascarids, with broad spectrum effects against other intestinal helminths also. Their principal problem is organophosphate-induced adverse reactions due to cholinesterase inhibition. Contraindications in general are any respiratory diseases, parturition within 30 days, evidence of diarrhea or other GIT problems, and the use or contemplated use of insecticides, muscle relaxants, phenothiazine-derived tranquilizers or CNS depressants. Organophosphates inhibit many enzymes, especially acetylcholinesterase, by phosphorylating its esterification site. This has the effect of blocking cholinergic nerve transmission in the parasite, resulting in spastic paralysis. The cholinesterases of man, animal hosts and parasites vary in their susceptibility to organophosphorus drugs. Because of such variations in susceptibility, attempts have been made to produce organophosphate products with maximum effects against the parasite, and minimum toxicity to the host and handler. In the host animal, the susceptibility of its cholinesterase enzymes to the organophosphate, the rate at which the inhibition can be reversed and also the rate of inactivation of the various organophosphate compounds largely determine their relative toxicity to different species.

The organophosphates tend to be labile to varying extents in alkaline media and may be partially hydrolysed and inactivated in the alkaline region of the small intestine. Thus, for example, the oral dose rate for trichlorfon (metrifonate) in cattle is 4.5 times that when given subcutaneously. In ruminants, the organophosphates generally have satisfactory efficacy for nematode parasites of the abomasum (especially *Haemonchus*) and small intestine, but lack satisfactory efficacy for parasites of the large bowel. Organophosphates are usually rapidly oxidized and inactivated in the liver. Their margin of safety is generally less than that of the other broad spectrum anthelmintics (benzimidazoles), and thus strict attention to dosage is necessary.

Dichlorvos is a particularly versatile organophosphate because of its high volatility. It can be incorporated as a plasticizer in vinyl resin pellets. The volatile dichlorvos is released slowly from the undigestible pellets as they pass through the digestive tract. This allows for a therapeutic concentration against parasites all along the digestive tract. This controlled release governs the concentration of dichlorvos available to the host and parasite, and also increases the safety margin by enabling the host animal to detoxify small quantities of the drug rather than being exposed to a sudden concentrated dose when not formulated in a resin. When passed in the feces, the pellets still contain approximately 45–50% of the original drug. Dichlorvos is rapidly absorbed and metabolized in the animal body.

Bound residues (indicated by prolonged radioactivity) are found in the tissues of some animal species after administration of dichlorvos. However, this persistent radioactivity is believed to be associated with the incorporation of the radiolabel into endogenous metabolism. When ^{14}C-vinyl-dichlorvos is orally administered to rats, the compound is rapidly metabolized. At 4 days, 40% is eliminated in expired air (CO_2) and 12% in urine (hippuric acid and urea). At 5 days, 75% of liver radioactivity is nonextractable material; ^{14}C is incorporated in endogenous compounds such as glycol, serine and cysteine (derived from CO_2) and this in proteins (Hutson et al., 1971).

9.6 TETRA HYDROPYRIMIDINES

Pyrantel was first introduced as a broad spectrum anthelmintic in 1966, initially for use against gastrointestinal parasites of sheep. It subsequently has come into use for cattle, horses, dogs and pigs. It is prepared for use as a tartrate embonate, or pamoate salt. Pyrantel tartrate is well absorbed in the pig and dog. There is less absorption of the drug by ruminants. Metabolism is rapid, with the metabolites being excreted in the urine (40% of the dose in the dog) and some unchanged drug being excreted in the feces (principally in the ruminant). Peak blood levels are usually attained 4–6 hours after oral administration. The pamoate salt of pyrantel is poorly soluble in water, which offers the advantage of reduced absorption of this particular salt from the gut, allowing the drug to reach and be effective against parasites in the lower end of the large intestine. Like levamisole, excretion is rapid. Pyrantel for oral use is usually in the form of a suspension, paste or drench, hence its application for human, equine and canine use. Aqueous solutions are subject to photoisomerization upon exposure to light, with a resultant loss in potency. Suspensions should therefore be kept out of direct sunlight. Pyrantel is not recommended for use in severely debilitated animals because of its levamisole-type phamacological action. Pyrantel is effective against ascarids, large and small strongyles, and pinworms.

Morantel is the methyl ester analogue of pyrantel, and in ruminant animals morantel tends to be somewhat safer and anthelmintically more effective than pyrantel. The drug is absorbed rapidly from the abomasum and upper small intestine of sheep. The drug is rapidly metabolized in the liver, and about 17% of the initial dose is excreted in the urine as metabolites within 96 hours after dosing.

Oral dosing of sheep with morantel tartrate (10 mg/kg) resulted in peak liver residue 1.92 ppm 48 hours after administration. Lowest liver residues, 0.27 ppm, occurred at 216 hours, after which time they were no longer detectable (Palermo, 1979). No residues were detectable in muscle tissue of cattle 14 days after dosing with a morantel drench (sensitivity of method 0.01 ppm) (Pompa, 1978).

Both pyrantel and morantel are most efficient against adult gut worms and larval stages which dwell in the lumen or on the mucosal surface. Activity is much less against the stages found in the mucosa, and is low against arrested osteragia larvae. A sustained-release ruminal bolus of morantel has recently been introduced for use in cattle, which acts as a controlled drug delivery system, releasing morantel over a 90-day period. This appears to be quite successful in preventing buildup of infective larvae on pasture, and thus the development of parasitic gastroenteritis and lungworm infestation.

The sustained release of morantel bolus effects a sustained delivery of active agent into the gut lumen. Relatively little morantel is absorbed into the systemic tissues from the bowel, and so the bolus has been cleared for use with a zero withholding time. That fraction which is absorbed is very rapidly metabolized. A new biodegradable bolus utilizing a polymeric matrix has been introduced in a number of countries.

9.7 AVERMECTINS

The avermectins, which are isolated from the mycelia of *Streptomyces avermitilis,* are a new family of antiparasitic compounds which are potent broad-spectrum agents at very low dosage levels. Since the original finding that 0.002 % of the avermectin complex in the diet completely cured mice infested with specific gastrointestinal nematodes, activity has been demonstrated against a wide range of nematode and arthropod parasites.

Avermectins are active when given orally, topically or parenterally, at dosages of a fraction of a milligram per kilogram, against many immature and mature nematode and arthropod parasites of sheep, cattle, dogs, horses and swine. The avermectins appear to paralyze nematodes and arthropods in a unique manner. Although there are a number of naturally occurring avermectins arising as fermentation products of the actinomycete, the derivative of most interest is avermectin B_1, a chemically modified derivative known as ivermectin. This particular compound is active against arrested and developing larvae and adults of the important cattle and sheep nematodes. Avermectins are naturally occurring fermentation products elaborated by the morphologically distinct soil organism *Streptomyces avermitilis* (Campbell et al., 1983). Early studies on the broth culture from this mold indicated the presence of a substance of unparalleled anthelmintic potency. Furthermore, this new substance displayed very high potency and high safety in the crude form as it occurred in nature, and without chemical modification. Early chromatographic studies yielded four separate entities, and on thin-layer chromatography these were seen to represent compounds with varying degrees of anthelmintic efficacy. The complex was seen to contain four components designated A_1, A_2, B_1, B_2 in varying proportions. These major components existed as two variants designated *a* and *b*. The *b* series was the lower homologue of the corresponding major *a* component (i.e., A_{1a} more potent than A_{1b}). Component B_{1a} was outstanding, being active against nematodes in sheep at a dosage rate of 0.025 mg/kg. The basic chemical structure of the avermectins is that of a macrocyclic lactone with two sugars attached (Burg et al., 1979). Removal of the sugar fraction from the molecule results in a marked diminution in potency, and in the anthelmintic activity. Analogs have been developed chemically which can be compared to the parent compound in activity and safety. The derivative of most interest in this context is 22.23 dihydroavermectin B_1, or ivermectin. This compound displayed good characteristics of efficacy and safety in the early laboratory and in vivo studies, and was, therefore, selected for further study and development (Chabala et al., 1980). Ivermectin comprises at least 80% 22.23 dihydroavermectin B_{1a} and not more than 20% of the B_{1b} homologue.

Ivermectin is teratogenic in rats, rabbits and mice but only at or near maternotoxic dose levels. Mice are the most sensitive species at a dosage of 0.2–0.4 mg/kg/day. Reproduction and multigeneration studies have demonstrated that neonatal rats are more susceptible to the toxic effects of ivermectin than adult rats. From mouse teratogenicity studies and from multigeneration studies in rats the NEL of 0.2 mg/kg/day has been established (Nessel et al., 1983).

Gamma-amino butyric acid (GABA) is the neurotransmitter substance mediating transmission of inhibitory signals from the interneurons to the motor neurons in the ventral nerve cord of the parasites. It is now established that ivermectin acts as a GABA agonist (Fritz et al., 1979; Melin et al., 1983). The function of this GABA transmitter is to open the chloride channels on the post synaptic function, allowing inflow of Cl^- ions and the induction of the resting potential. Ivermectin potentiates this effect by stimulating the presynaptic release of GABA, and by increasing its binding to the postsynaptic receptors. In the presence of ivermectin, the chloride channels are open when they should be closed, the net effect being that signals and impulses are not received by the recipient cell. Although the motor neuron and muscle cells are both capable of individual excitation, passage of the electrical impulse across the synapse is blocked. Arthropods utilize GABA as a neurotransmitter but tend not to utilize it between two sets of nerve cells, as in nematodes, but between nerve and muscle cells. Prolonged stimulation of GABA release renders the effects of ivermectin sustained and irreversible (Egerton et al., 1980). For most parasites, this results in neuromuscular blockade, paralysis and death.

The overall GABA-mediated chloride ion conductance effect may be due to

- ivermectin acting as a GABA agonist either at the GABA binding site or elsewhere on the protein
- stimulation of presynaptic GABA release
- potentiation of GABA binding to its receptors

In experimental work it was further observed that washing neurons with picrotoxin (an antagonist of GABA) abolished this ivermectin-induced paralysis.

Paralysis is the most evident effect of ivermectin in parasites, but suppression of reproductive function has also been observed in ticks. Ivermectin displays no activity against cestodes or trematodes, as these parasites do not utilize GABA as a neurotransmitter. This is consistent with the hypothesis regarding mode of action.

Ivermectin is well absorbed when administered orally or parenterally. The route of administration and the formulation employed affects its disposition profile. High levels of ivermectin are reached in the lungs and skin regardless of the route of administration. Efficacies greater than 95% have been achieved against parasites of the skin, respiratory system, and blood after oral treatment.

Concentrations of ivermectin are maintained in body fluids for prolonged periods of time (Bennett, 1986). In cattle dosed subcutaneously with ivermectin at 0.3 mg/kg, a half-life of 70 hours per total radioactive residue in plasma was reported (Jacob et al., 1983). Following intravenous delivery of 0.2 mg/kg ivermectin in sheep, a terminal half-life of 178 hours was detected (Prichard et al., 1985). This relatively long half-life is related to the very high potency of the compound, as studies with other anthelmintics have indicated that efficacy is profoundly affected by the kinetic profile. In sheep, low bioavailability was reported when ivermectin was administered into the rumen. Some degradation in the rumen may account for this reduced bioavailability. The monosaccharide and aglycon derivatives (two possible metabolites of rumenal degradation) are less potent than the parent drug. The lower efficacy and shorter duration of action of orally administered ivermectin (compared with parenteral dosing) may explain the less potent effect against certain ticks *(Boophilus sp.)* in cattle and body lice in sheep. There is ample evidence of the prolonged half-life of ivermectin, based on antiparasitic trials. Cattle did not become infected when exposed to infective stages of gastrointestinal parasites, lungworms, and mange mites 14 to 21 days after SC treatment. Blood-sucking lice and flies died after feeding on cattle treated 10 to 14 days earlier.

Most of the administered dose of ivermectin is excreted in the feces, the remainder in the urine (Campbell et al., 1983). Minimal residues are present in the muscle and kidneys, with the highest concentrations being detected in the liver and fat tissues. Residues in all tissues are extractable in nature with little or no macromolecularly bound drug or metabolites present. The major single component in the edible tissues of cattle, sheep and pigs is the unaltered parent drug. Ivermectin is also excreted through the milk of lactating cattle.

Although mammals utilize gamma-aminobutyric acid as a central neurotransmitter, they are generally not adversely affected by ivermectin. This is because being a macrolide of large molecular weight, ivermectin does not readily cross the blood-brain barrier of the mammal to affect the GABA within the central nervous system.

In tests of brain concentration of the drug in cattle, radioactive residue assays revealed only minute traces of ivermectin. This was the lowest concentration of all tissues analyzed. There have been a number of cases of central nervous system depression in purebred and crossbred long-haired collies. The reason for this breed susceptibility is not known. It has been postulated that the blood-brain barrier in the collie may be more permeable to ivermectin than in other species, allowing ivermectin to enter the central nervous system.

Studies on the metabolic disposition of ivermectin have been carried out in cattle, sheep, swine, and rats, with the drug labelled with tritium in the C-5 or C-22 and C-23 positions. The animals were dosed once at levels of 0.3 to 0.4 mg per kilogram of body weight by the subcutaneous, oral, or intraruminal route. Regardless of the route only 0.5 to 2.0 percent of the administered radioactivity was excreted in the urine. The remainder appeared in the feces (Campbell et al., 1983).

Animals were slaughtered over a period of 1–28 days after treatment, and about 25 tissues and body fluids were assayed for total radioactive residues. The liver and fat contained the highest radioactive residues in all species, with very little residue in the muscle and kidney. The radioactive residue in the edible tissues of cattle, sheep, and swine, as well as in plasma, was essentially all extractable in organic solvents such as toluene or methylene chloride. Thus, there were very little, if any, intractable, macromolecularly bound drug or metabolite residues. Solvent partition followed by reversed-phase high performance liquid chromatography (HPLC) proved extremely effective in separating the components of the tissue residues. Analysis of these tissues showed that the major single component in the edible tissues of the cattle, sheep, and swine was the unaltered drug (Campbell et al., 1983).

An unusually nonpolar fraction consisting of several HPLC peaks, and accounting for about 60% of the total residue in steer fat collected 28 days after treatment, was isolated by solvent fractionation and HPLC. The major hydrolysis product was identified as the same 24-hydroxymethyl metabolite found in steer liver. These compounds were probably present in fat as the acyl esters of the metabolite. The metabolites found in the edible sheep tissues were very much like those found in the steer, and the metabolic disposition of the drug in the rat was very similar to that in the other species.

So that the tissue distribution of ivermectin residues in cattle and sheep could be studied, and the requirements of governmental regulatory agencies could be satisfied, a sensitive, specific, and reliable chemical assay for the parent drug and any major drug-related metabolites was needed. Radioactive metabolism studies have shown that dihydroavermectin B_{1a} is the major residue found in all tissues prior to the proposed withdrawal times of 28 days for cattle and 14 days for sheep. Dihydroavermectin B_{1b} is dosed at lower levels than dihydroavermectin B_{1a} and is generally metabolized more rapidly than the B_{1a} component, so that the dihydroavermectin B_{1b} residues are always less than the B_{1a} residues. Ivermectin is such a potent drug, and is dosed at such low levels, that the tissue assay needs a sensitivity of 10 ppb (10 mcg/kg) to ensure negligible residues.

An analytical procedure has been developed for the isolation and determination of ivermectin in sheep and cattle tissues. The method is based upon the detection of a fluorescent derivative of this compound following high-performance liquid chromatography. The assay has a lowest limit of reliable measurement of 10 ppb and a limit of detection of 1–2 ppb. Recoveries of ivermectin spikes average 83% for liver, kidney, muscle and fat. In residue studies, liver appears to retain the residue the longest; a negligible residue of 11 ppb was found in cattle liver at the proposed withdrawal time of 28 days. In sheep liver, no residue was detected at the proposed withdrawal time of 14 days (Tway et al., 1981).

A residue study was carried out on cattle dosed at 0.3 mg/kg with ivermectin by subcutaneous injection. The liver, which is the tissue in which the drug is most persistent, had average residues ranging from 454 ppb at 2 days post-dose to 11 ppb at 28 days post-dose. The other tissues (fat, muscle, and kidney) had lower residues than liver at all withdrawal times in the order fat > kidney > muscle. No residues were detected in tissues from the control and 42-day animals. In muscle, kidney, and fat, peak drug levels were reached between 2 and 7 days post-dose. At withdrawal times of 7 days and beyond, the drug residue in these tissues decreased linearly as a logarithmic function. The drug half-lives in muscle, kidney, and fat were 3.3, 3.9, and 3.7 days, respectively. Not only are the highest residue levels found in liver but also the drug has the longest half-life in liver. Liver is the last tissue to clear drug-related residues, but all tissues have no detectable residues by 42 days post-dose.

A residue study was also done on sheep dosed with oral ivermectin at 0.3 mg/kg. Peak drug levels were found in all tissues at 1 day post-dose. At 1 and 3 days post-dose, fat had higher residues than liver, but by 5 days post-dose, liver and fat had equivalent residues. At 5 and 7 days post-dose, the drug residue in both fat and liver were at the assay's limit of reliable measurement, so that analytically there is no difference between residue values of 8 and 9 ppb. At 10 and 14 days post-dose, no tissue contained any detectable residue (Tway et al., 1981).

Based on the relationship of the NEL to residue levels in edible tissues from pharmacokinetic studies, the current withdrawal times for sheep and cattle have been established. However, on account of its very high potency and elimination through milk, ivermectin preparations are contraindicated for use in lactating cattle producing milk for human consumption. Year round parasite control programs can involve 3–4 treatments per year in young stock. The primary impact on the environment from the use of ivermectin in farm animals is the excretion of the drug by treated animals via their feces and urine.

Ivermectin in feces/soil degrades at a slow but significant rate. In a winter environment, decomposition is slow, necessitating a period of 13 weeks for ivermectin levels to deplete by 50 percent. When exposed to an outdoor summer environment ivermectin in soil decomposes with a half-life of 0.8 weeks.

Ivermectin possesses little adverse effect on fresh water algae, but only limited information is available on the application of the drug to plant foliage or roots. Studies with the dung beetle have shown that while ivermectine residues in feces from treated animals does not adversely affect the adult beetles, larvae in the treated dung do not complete their development. Larval development of the buffalo fly in fresh feces of treated cattle is prevented for up to 14 days after treatment.

The concentration of ivermectin and metabolites in the accumulated waste of cattle and sheep is approximately 18–19 ppb. Assuming normal water-spreading practices are observed on pasture, concentration of dung and metabolites in soil is estimated to be less than 0.1 ppb. From laboratory experiments, exposure levels of 300 mg/kg of soil (ppm) are necessary for toxicity to earthworms. Thus a margin of 3 million exists between anticipated pasture/soil levels and the toxic levels for earthworms (Nessel et al., 1983).

The use of ivermectin ruminal boluses in calves has recently been shown to affect dung degradation in implanted animals (Wall, 1987). Dung samples collected from calves fitted with ruminal boluses delivering ivermectin at 40 μg per kg per day contained a total of only 17 adult dung beetles, 35 dipterous larvae and 44 earthworms after 100 days compared with control which yielded 780 dung beetles, 267 dipterous larvae and 46 earthworms over the same period. Although these dung-induced effects may vary with the manner in which the drug is administered as well as with the geographical location and climate, the routine use of ivermectin in cattle may pose an ecological threat to those insects (particularly diptera) that have evolved to exist uniquely in conjunction with cattle dung (Wall and Strong, 1987). Failure of dung degradation may be attributable to the absence of insects which are eliminated by the insecticidal effects of ivermectin. Widespread use of ivermectin may have important environmental consequences for pastureland in certain geographical locations. Abamectin for plant protection is currently being introduced in a number of countries.

9.8 SUMMARY

A considerable additional volume of information is necessary to properly evaluate toxic hazards of benzimidazole residues. Residues are highest and last longest in the liver. Withdrawal times can be established on the basis of total residue depletion, extractable metabolite toxicological studies or the sum of potentially teratogenic metabolites. Because of the bio-interconversion of a number of metabolites, residue studies must consider total residues rather than any single metabolite. Benzimidazole compounds have been shown experimentally to induce mitotic arrest, chromosome doubling, and possess mutagenic capability. Genotoxicity and carcinogenicity potential must be closely examined for benzimidazole anthelmintics. Present evidence suggests that because of the low residue levels in target animal tissues, and the observed low bioavailability, covalently bound residues should not present a major hazard. In the light of current scientific knowledge, their toxicological significance is speculative. Current information would indicate that residue levels of benzimidazoles are unlikely to pose a health hazard to consumers when approved withdrawal times are rigidly observed. For the present, only provisional ADI's can be established for benzimidazoles regarding flukicides. On account of their narrow safety margins, pharmacodynamic effects, long half-lives, high binding to plasma protein, deposition in the liver and their excretion in detectable quantities in milk, it is imperative that dosage and withdrawal times are observed. Because these fasciolicides are used no more than 2 or 3 times per year, residues are unlikely to occur in milk with any great frequency. Nonetheless the ADI could possibly be exceeded in the case of infants or small children, consuming one liter of milk per day, and with a body weight of 10 kg or less, even at the end of the withholding period.

Ivermectin is widely used in food-producing animals for the control of various parasitic diseases. For the most part, like many other parasiticides it is not subject to legislative controls, and so is freely available to the layman. It is now available in pour-on form in some countries. This could pose a hazard to the handler by way of ingestion, careless use, contact exposure, etc. Possessing a unique mode of action, ivermectin is characterized by a very slow depletion time from body tissues, a persistent clinical effect, excretion through feces and milk and very high potency against parasites of the skin following oral, topical or parenteral administration of minute doses. A slow release bolus of ivermectin is currently being developed for even greater persistent effects. While current data would suggest that residues presented to the consumer at the end of the recommended withholding period are negligible and of minimal risk, there is a need for further toxicological studies and development of improved analytical techniques to further characterize the kinetic profile and to properly evaluate hazards arising from ingestion of milk from treated animals and the risks to the environment. It has been used widely in human medicine for the treatment of *Onchocerca* infestation without appreciable hazard.

Organophosphates can provide hazards to man. Being lipid soluble they are well absorbed through the unbroken skin. Cholinesterase inhibition can occur in man following accidental organophosphate exposure. These drugs also have the propensity towards interaction with many other drugs. Although relatively rapidly degraded in the animal body, tissue residues are unlikely to pose a serious consumer hazard if specified withholding times are enforced. Perhaps the major risk arises from contamination of the environment, from fecal excretion or accidental drug spillage. Fish are particularly susceptible to organophosphate poisoning, and many instances of serious water pollution and fish kills have been attributed to careless disposal of organophosphate pesticides. Sprays, collars and washes of organophosphates used in small animals can present significant hazards to young infants following ingestion, inhalation or transcutaneous absorption of the active principle.

9.9 REFERENCES

Anon. 1983. "Anthelmintics for Cattle, Sheep, Pigs and Horses," Ministry of Agriculture, Fisheries and Food, London.

Averkin, E. A., C. C. Beard, C. A. Dvorak, J. A. Edward, J. H. Fried and J. G. Kilian. 1975. "Methyl 5(6)-phenylsulfinyl-2-benzimidazolecarbamate, a New Potent Anthelmintic," *Journal of Medicinal Chemistry*, 11:1165–1166.

Baer, J. E., T. A. Jacobs and F. J. Wolf. 1977. *J. Toxicol. Environ. Health*, 2:895–903.

Becker, W. 1975. *Proc. Eur. Multicolloq. Parasitol. 2nd Trogir*, Yugoslavia, pp. 105–106.

Bennett, D. G. 1986. *J. Am. Vet. Med. Assoc,* 1:100–104.

Berger, H., T. R. Garces, R. K. Fisher, et al. 1987. *Am. J. Vet. Res.,* 48(5):852–854.

Bogan, J. A., J. Armour, K. Bairden and E. A. Galbraith. 1987a. "Time of Release of Oxfendazole from an Oxfendazole Pulse-Release Bolus," *Veterinary Record,* 121:280.

Bogan, J. A. and S. E. Marriner. 1983. "Pharmacokinetics of Albendazole, Fenbendazole and Oxfendazole," in *Veterinary Pharmacology and Toxicology,* Y. Ruckebusch, P. Toutain and G. D. Koritz, eds., M.T.P. Press Ltd., pp. 235–239.

Bogan, J. A. and S. E. Marriner. 1982. *Pharmacol Toxicol. Veter.* Paris: INRA Public, p. 313.

Bogan, J. A. and S. E. Marriner. 1984. "Pharmacodynamic and Toxicological Aspects of Albendazole in Man and Animals," in *Albendazole in Helminthiasis,* M. Firth, ed., RSM Int. Cong. Sym. Ser. 61. RSMB.

Bogan, J. A., S. E. Marriner and E. A. Galbraith. 1982. *Res. Vet. Sci.,* 32:124–126.

Boyd, M. R. and L. T. Burka. 1978. "In Vivo Studies on the Relationship Between Target Organ Alkylation and the Pulminary Toxicity of a Chemically Reactive Metabolite of 4-Ipomeanol," *Journal of Pharmacology and Experimental Therapeutics,* 207:687–697.

Burg, R. W., B. M. Miller, E. E. Baker, J. Birnbaum, S. A. Currie, R. Hartman, Y. L. Kong, R. L. Monaghan, G. Olson, I. Putter, J. D. Tunac, H. Wallack, E. O. Stapley, R. Oiwa and S. Omura. 1979. *Antimicro. Agents Chemother.* 15:361–367.

Burgat-Sacaze, V., P. Delatour and A. Rico. 1981. *Ann. Rech. Vet.,* 13(3): 277–289.

Campbell, W. C., M. H. Fisher, E. O. Stapley, G. Albers-Schonberg and T. A. Jacob. 1983. *Science,* 221:823–828.

Chabala, J. C., H. Mrozich, R. L. Tolman, R. L. Eskola, A. Luis, L. H. Peterson, M. F. Woods, M. M. Fisher, W. C. Campbell, J. R. Egerton and D. A. Ostlind. 1980. *J. Med. Chem.,* 23:1134–1136.

Chement, D. 1982. "Essal de Traitement des Strongylidoses Digestives des Equides par le Febantel," These Doct. Vet. Paris.

Coles, G. C. and N. G. Brisco. 1978. "Benzimidazoles and Fluke Eggs," *Veterinary Record,* 103:301–361.

De Caprio, A. P., E. Olajoe and P. Weber. 1982. "Covalent Binding of a Neurotoxic N-Hexane Metabolite: Conversion of Primary Amines to Substituted Pyrole Adducts by 2,5-Hexadione," *Toxicology and Applied Pharmacology,* 65:440–450.

Delatour, P. M. Caudon, F. Carnier and E. Benoit. 1982. "Relationship of Metabolism and Embryotoxicity of Febantel in the Rat and Sheep," *Annales Recherch Veterinaire,* 13:163–179.

Delatour, P., M. C. Cure, E. Benoit and F. Garnier. 1986. *J. Vet. Pharmacol. Ther.,* 9:230–235.

Delatour, P., M. Yoshimura, F. Garnier and E. T. Benoit. 1982. *Rec. Med. Vet.,* 158:369–373.

Delatour, P., M. Daudon, F. Garnier and E. Benoit. 1982. *Ann. Rech. Vet.,* 13:163–170.

Delatour, P., M. P. Tiberghien, F. Garnier and E. Benoit. 1985. *Am. J. Vet. Res.,* 46:1399–1402.

Delatour, P., G. Lorgue, M. Lapras, Y. Richard, M. C. Nain and G. Cantenot. 1975. *Bull. Soc. Sci. Vet. Med. Comp. Lyon.,* 17:185–194.

Delatour, P., G. Lorgue, D. Courtot and M. Lapras. 1975. *Bull. Soc. Sci. Vet. Med. Comp. Lyon.,* 77:197–203.

Delatour, P., J. Debroye, G. Lorgue and D. Courtot. 1977. *Rec. Med. Vet.*, 153:639–645.

Delatour, P., M. P. Tiberghien and S. Besse. 1983. *J. Vet. Pharmacol. Therap.*, 6:233.

Delatour, P., D. Longin, F. Garnier and E. Benoit. 1983. *Ann. Rech. Vet.*, 14:65–70.

Delatour, P. and R. Parish. 1986. *Benzimidazole Anthelmintics and Related Compounds: Toxicity and Evaluation of Residues*, A. G. Rico, ed., Acad. Press, pp. 175–204.

Delatour, P., R. C. Parish and R. K. Gyurik. 1981. *Ann. Tech. Vet.*, 12:159.

Delatour, P., M. Daudon, F. Garnier and E. Benoit. 1983. *Ann. Rech. Vet.*, 13:163.

Dicuollo, C. J., J. A. Miller, W. L. Mendelsohn and J. F. Pagano. 1974. *J. Agr. Food Chem.*, 22(6):948–953.

Duwel, D. 1977. "Fenbendazole. II. Biological Properties and Activity," *Pesticide Science*, 8:550–555.

Egerton, J. R., J. Birnbaum, L. S. Blair, J. C. Chabala, J. Conroy, M. H. Fisher, H. Mrozik, D. A. Ostlind, C. A. Wilkins and W. C. Campbell. 1980. *Br. Vet. J.*, 136:88–97.

Friedman, P. A. and E. G. Platzer. 1978a. *Biochem. Biophys. Acta.*, 544:605–614.

Friedman, P. A. and E. G. Platzer. 1978b. "Interaction of Anthelmintic Benzimidazoles and Benzimidazole Derivatives with Bovine Brain Tubulin," *Biochemica et Biophysica Acta.*, 544:605–614.

Fritz, L. C., C. C. Wang and A. Gorio. 1979. *Proc. Natl. Acad. Sci.*, 76:2062–2066.

Gordon, G. B., S. P. Spielber, D. A. Blake and V. Balasubramanian. 1981. "Thalidomide Teratogenesis: Evidence for a Toxic Arene Oxide Metabolite," *Proceedings of the National Academy of Sciences, USA*, 78:2545–2548.

Granned, C. 1980. "Albendazole: Embryotoxicite de Relais," These Doctorat Veterinaire, Ecole de Medicine Veterinaire, Lyon.

Gyurik, R. J., A. W. Chow, B. Zaber, et al. 1981. *Drug Metab. and Dispos.*, 9(6):503–507.

Havercroft, J. C., R. A. Quinian and K. Gull. 1981. *J. Cell. Soc.*, 49:195–204.

Heeschen, W. 1979. *I.D.F. Bulletin, Document 113*, pp. 30–39.

Hennessy, D. R. 1985. "Examination and Manipulation of the Pharmacokinetic Behavior of Benzimidazole Anthelmintics," Ph.D. Thesis, Marquarie University, Sydney.

Hennessy, D. R. and R. K. Prichard. 1981. *Proc. Aust. Soc. Parasitology*, Perth.

Hoebeke, J., G. Van Nijen and M. de Brabander. 1976. *Biochem. Biophys. Res. Commun.*, 69:319–324.

Hutson, D. H., E. C. Hoadley and B. A. Pickering. 1971. *Xenobiotica*, 1:593–611.

Ireland, C. M., L. Clayton, W. E. Gutteridge, C. I. Robson and K. Gull. 1982. "Identification and Drug Binding Capabilities of Tubulin in Nematode *Ascaridia guli*," *Molecular Biochemical Parisitology*, 6:45–53.

Ireland, C. M., K. Gull, W. E. Gutteridge and C. I. Pogson. 1979. *Biochem. Pharmacol.*, 28:2680–2682.

Irons, R. D., J. G. Dent, T. S. Baker and D. E. Rickert. 1980. "Benzene Is Metabolised and Covalently Bound in Bone Marrow in Situ," *Chemistry-Biology Interactions*, 30:241–245.

Jacob, T. A., R. P. Buhs, J. R. Carlin, S. H. L. Chiv, G. Miwa and A. Rosegay. 1983.

"The Metabolism and Tissue Residue Profiles of Ivermectin," in *Recent Developments in the Control of Animal Parasites, Proc. MSD-Agvet Symp.*, W. H. D. Leaning, ed., Perth Australia, pp. 56–97.

Jallow, D. J., J. R. Mitchell, W. Z. Potter, D. A. Davis, J. R. Gillette and B. B. Brodie. 1973. "Acetaminophen-Induced Hepatic Necrosis. II. Role of Covalent Binding in Vivo," *The Journal of Pharmacology and Experimental Therapeutics*, 187: 195–202.

Johns, D. J. and J. R. Phillip. 1977. *Proc. Int. Conf. Word Ass. Adv. Vet. Parasitol. 8th.* Sydney.

Juszkiewicz, T. et al. 1971. *Europ. J. Toxicol.*, 46:525.

Kappas, A., M. H. L. Green, B. A. Bridges, A. M. Rogers and W. J. Muriel. 1976. *Mutagen Res.*, 40:379–382.

Kelly, J. D. and C. A. Hall. 1979. "Resistance of Animal Helminths to Anthelmintics," *Advances in Pharmacology and Chemotherapy*, 16:89–128.

Kohler, P. and R. Bachman. 1981. "Intestinal Tubulin as Possible Target for Chemotherapeutic Action of Mebendazole in Parasitic Nematodes," *Molecular Biochemica Parasitology*, 4:325–336.

Lapras, M., P. Delatour, J. Labe, M. J. Panarin and Y. Richard. 1975. *Bull. Soc. Sci. Vet. Med. Comp. Lyon*, 77:379–397.

Lapras, M., J. P. Deschanel, P. Delatour, J. Gastellu and M. Lombard. 1973. *Bull. Soc. Sci. Vet. Med. Comp. Lyon*, 75:53–61.

Lee, R. 1973. *Parasitology*, 67:xiii.

Marriner, S. E. and J. A. Bogan. 1981. *Am. J. Vet. Res.*, 42:1146–1148.

Marriner, S. E. and J. A. Bogan. 1981a. *Am. J. Vet. Res.*, 42:1143–1145.

Marriner, S. E. and J. A. Bogan. 1980. *Am. J. Vet. Res.*, 41:1126–1129.

Martz, F., C. Failinger III, and D. A. Blake. 1977. "Phenytoin Teratogenesis: Correlation Between Embryopathic Effect and Covalent Binding of Putative Arene Oxide Metabolite in Gestational Tissue," *Journal of Pharmacology and Experimental Therapeutics*, 203:231–239.

McDougall, J., B. D. Cameron and D. Lichtenwainer. 1985. *Congr. Eur. Ass. Vet. Pharmacol. Toxicol.*, 3rd Ghent, Belgium.

Mellin, T., R. D. Busch and C. C. Wang. 1983. *Neuropharmacology*, 22:89–96.

Miller, J. A. and E. C. Miller. 1976. "The Metabolic Activation of Chemical Carcinogens. Recent Results with Aromatic Amines, Safrole and Aflatoxin B_1," *I.A.R.C. Scientific Publications*, 12:153–170.

Mitchell, J. P., W. Z. Potter and D. J. Lollow. 1973. "Furosemide-Induced Hepatic and Renal Tubular Necrosis. 1. Effects of Treatments Which Alter Drug Metabolizing Enzymes," *Federation Proceedings*, 32:305.

Mochida, K., M. Goto and K. Saito. 1983. *Bull. Environ. Contam. Toxicol.*, 31:428–431.

Mohammed-Ali, N. A. K. and J. A. Bogan. 1987. *J. Vet. Pharm. Therap.*, 10:127–133.

Morris, D. L., P. W. Sykes, B. Dickson, S. E. Marriner, J. A. Bogan and F. G. O. Burrows. 1983. "Albendazole in Hydatid Disease," *British Medical Journal*, 286:103–104.

Nerenberg, C., R. A. Runken and S. B. Martin. 1978. "Radioimmunoassay of Oxfendazole in Bovine, Equine or Canine Plasma or Serum," *Journal of Pharmaceutical Science*, 67:1553–1557.

Nessel, R. J., T. A. Jacob and R. T. Robertson. 1983. "Human and Environmental Safety Aspects of Ivermectin," in *Recent Developments in the Control of Animal Parasites, Proc. MSD-Agvet Symp.*, Perth, Australia, W. H. D. Leaning, ed., pp. 98–108.

Ngomuo, A. J., S. E. Marriner and J. A. Bogan. 1984. "The Pharmacokinetics of Fenbendazole and Oxfendazole in Cattle," *Veterinary Research Communications,* 8:187–193.

Ngomuo, A. J. 1984. "Pharmacokinetic Studies of Fenbendazole and Oxfendazole in Cattle," MVM Thesis, Univ. of Glasgow.

Palermo, D. 1979. *Archivio Vet. Ital. Della Sci. Vet.*,30(3/4):110–111.

Parker, S. W. 1982. "Allergic Reactions in Man," *Pharmacological Reviews,* 34:85–104.

Pompa, G. 1978. *Atti Della Soc. Ital. Della Sci. Vet.*, pp. 89–95.

Prichard, R. K. 1978. "Anthelmintics," In *Proceedings of 39th Refresher Course for Veterinarians.* Univ. of Sydney, 6:421–463.

Prichard, R. K. 1973. "The Fumarate Reductase Reaction of *Haemonchus contortus* and the Mode of Action of Some Anthelmintics," *International Journal of Parasitology,* 3:409–417.

Prichard, R. K., J. W. Steel, E. Lacey and D. R. Hennessey. 1985. *J. Vet. Pharmacol. Ther.*, 21:87–94.

Prichard, R. K., D. R. Hennessey and J. W. Steel. 1978. *Vet. Parasitol,* 4:309–315.

Rew, R. S. 1978. *J. Vet. Pharmacol. Ther.*, 1:183–198.

Rosenblum, C., N. R. Trenner and D. E. Wolf. 1971. *J. Labelled Compd.*, 8:225–232.

Schreiner, C. A. and H. E. Holden. 1983. In *Teratogenesis and Reproductive Toxicology,* E. M. Johnson and D. M. Kockhar, eds., Berlin: Springer Verlag, pp. 135–167.

Scott, G. C. and S. J. Di Cuollo. 1980. *Int. Conf. Eur. Ass. Vet. Pharmacol. Toxicol.* Cambridge.

Seiler, J. P. 1973. *Experienta,* 29:622–623.

Short, C. R., S. A. Barker, L. C. Hsieh, S. P. Ou, L. E. Davis, G. Koritz, C. A. Neff-Davis, R. F. Bevill and I. J. Munsiff. 1987a. "The Disposition of Fenbendazole in the Goat," *American Journal of Veterinary Research,* 48:811–815.

Short, C. R., S. A. Barker, L. C. Hsieh, S. P. Ou, L. E. Davis, G. Koritz, C. A. Neff-Davis, R. F. Bevill and I. J. Munsiff. 1987b. "The Disposition of Fenbendazole in Cattle," *American Journal of Veterinary Research,* 48:958–961.

Simkins, K. L., E. J. Orioski, W. E. Schumacher, et al. 1972. *J. An. Sci.*, 35:193 (Abstr.).

Simkins, K. L., J. E. Smith and R. G. Eggert. 1975. *J. Dairy Sci.*, 59(8):1440–1443.

Steel, J. W., D. R. Hennessy and E. Lacey. 1985. *Abstr. Conf.*

Steel, J. W., D. R. Hennessy and E. Lacey. 1985. *Netobimin (Totabin-Sch): Metabolism and Pharmacokinetics in Communication to the Congress of the World Association for Advancement of Veterinary Parasitology,* Rio de Janeiro, p. 38.

Styles, J. A. and R. Garner. 1974. *Mutagen Res.*, 26:177–187.

Swenson, D. H., E. C. Miller and J. A. Miller. 1974. "Aflatoxin B_1-2,3-oxide: Evidence for Its Formation in Rat Liver in vivo and by Human Liver Microsomes in Vitro," *Biochemistry and Biophysics Research Communications,* 60:1036–1043.

Szabo, K. T., C. R. Miller and G. C. Scott. 1974. *Cornell Vet. 64. Suppl.* 4:41–55.

Theodorides, V. J., C. J. Di Cuollo, T. Nawalinski, C. R. Miller, J. R. Murphy, J. F. Freeman, J. C. Kileen and W. R. Rapp. 1977. *Am. J. Vet. Res.*, 38:809–814.

Til, H. P. and M. Williams. 1970. Central Insit. Voor Voedingsonderzoek, Amsterdam.

Tocco, D . J., J. R. Egerton, W. Bowers, V. W. Christensen and C. Rosenblum. 1965. *J. Pharmacol. Exp. Therap.*, 149:263–271.

Tocco, D. J., R. P. Buhs, H. D. Brown, A. R. Matzuk, H. E. Mertel, R. E. Harman and N. R. Trenner. 1964. *J. Med. Chem.*, 7:339.

Tsina, I. W. and S. B. Matin. 1981. *J. Pharmaceut. Sci.*, 70(8):858–860.

Tway, P. C., J. S. Woods and G. V. Downing. 1981. *J. Agric. Food Chem.*, 29:1059–1063.

Vanden Bossche, H., F. Rochette and C. Horig. 1982. *Adv. Pharmacol. Chemother.*, 19:67–128.

Vanden Bossche, H. and S. Nollin. 1973. *Int. J. Parasitol.*, 3:401–407.

Vander Meer, S., and Pouwels, M. 1969. *J. Med. Chem.*, 12:539.

Wall, R. and L. Strong. 1987. *Nature*, 327, 6121:418–421.

Watson, D. M. 1983. *Food Chem.*, 12:167.

Watts, S. A. M. 1981. "Colchicine Binding in the Rat Tapeworm, *Hymenolepis diminuta,*" *Biochemica et Biophysica Acta*, 667:59–67.

Weir, A. J. and J. A. Bogan. 1985. *J. Vet. Pharmacol. Therap.*, 8:413–414.

Wetzel, H. 1985. *Zentralbl. Vet. Med.*, 32:375–394.

CHAPTER 10

Pesticide Residues in Foods of Animal Origin

W. D. BLACK, D.V.M., M.Sc., Ph.D.[1]

10.1 INTRODUCTION

PESTICIDES became important tools in the production of food during the 1960s and 1970s. Usage has levelled off since the late 1970s (Kulkarni and Ashoke, 1990); however, expanded usage is projected in the future and well into the next century (Mehortra, 1986). The requirement of an increasing global population for food, as well as the need for a means of controlling parasites and diseases which threaten man's well-being and comfort, have been the driving forces (Mehrotra, 1986). Although designed and used to kill selected forms of life, their value to mankind is to selectively protect and foster the growth of important species. The term *pesticide* can be used to describe any chemical that kills or controls pests including weeds, insects, fungi, bacteria and rodents (Committee on Scientific and Regulatory Issues Underlying Pesticide Use Patterns and Agricultural Innovation, 1987a). For the purposes of this review the term will be limited to insecticides plus a limited number of other chemicals (e.g., PCB, PBB) which are detected in residue monitoring systems.

Although DDT was first thought to be the cure-all for pest problems affecting mankind, within 5 years of its introduction DDT-resistant house flies were detected (Busvine, 1976). As a result of the development of resistant insects and changing safety requirements, DDT is no longer available in the developed countries of the world and newer, safer chemicals are now available (Table 1). DDT and HCH (hexachlorohexane)

[1]Department of Biomedical Sciences, Ontario Veterinary College, University of Guelph, Guelph, Ontario, Canada.

TABLE 10.1. Insecticides Available for Use on Livestock and Livestock Premises in Canada (1991).

Application	Chemical	Insecticide	Livestock
Ear Tags	Pyrethrins	Permethrin	cattle*
		Cypermethrin	cattle*
		Fenvalerate	cattle*
		Flucythrinate	cattle*
	Organophosphates	Tetrachlorvinphos	cattle
Livestock Premises	Pyrethrin	Pyrethrins	barns and houses
	Metabolic Inhibitors	Piperonyl butoxide	barns and houses
	Carbamate	Methomyl	barns—cattle, swine, poultry
	Organophosphates	Dimethoate	barns and houses
		Dichlorovos	barns—cattle, swine, poultry
		Malathion	barns—cattle, swine, poultry
		DDVP	barns—cattle, swine, poultry
	Carbamates	Carbaryl	barns—cattle, swine, poultry
Topical Application	Pyrethrin	Pyrethrins	cattle, swine
		Permethrin	beef cattle, poultry
	Metabolic Inhibitors	Piperonyl butoxide	cattle, swine
	Organophosphates	Crotoxyphos	cattle
		Dichlorovos	cattle
		Fenthion	swine
		Tetrachlorvinphos	poultry
		Coumaphos	cattle, swine
	Organo-chlorines	Lindane	beef cattle, swine, sheep
	Derris	Rotenone	cattle, swine, sheep
		Trichlorfon	cattle
Oral and Injection	Avermectins	Ivermectin	cattle, sheep, swine

*Dairy and beef cattle, nonlactating dairy cattle included with beef.
Source: Canadian Compendium of Veterinary Pharmaceuticals, Biologicals and Specialties, 1991.

continue to be used in some tropical countries to control malaria, and problems with residues in these countries are widespread (Mehrotra, 1985a, 1985b).

By the late 1950s an unexpected problem was recognized. Rachel Carson first brought it to the attention of the general public in her novel *Silent Spring* in which she warned of an impending ecological disaster if use of pesticides such as DDT was not curtailed. The impact of this novel on the North American consciousness, along with the development of the environmental movement, had much to do with the regulatory actions which will be discussed below, and the changes which have occurred in the use of pesticides. Before the regulatory changes took effect, and before data on the stability and accumulation of these chemicals in the environment became known, fish and seals in wilderness areas thousands of kilometers from the sites of application were contaminated by DDT and its metabolite residues (WHO International Task Group on Environmental Health Criteria for DDT, 1989). Further investigation revealed residues of these agents in meat, milk and vegetables, and in human breast milk and human fat (Frank et al., 1970; Saxena and Siddiqui, 1982). Decisive steps were needed. These included the banning of some chemicals, restrictions on the use of others, and development of monitoring systems for pesticide residues in food and the environment. DDT was the first to be removed from agricultural use in the late 1960s, and by 1978 most of the organo-chlorines were no longer used in developed countries of the world.

10.2 IMPORTANCE OF PESTICIDES TO MAN

The use of pesticides on livestock was rather inefficient until the discovery of DDT [1,1,1-trichloro-2,2-bis(p-chlorophenyl) ethane] in 1939 by Muller working at the Basle Labs of Geigy (Lauger et al., 1944) and its subsequent widespread deployment in the 1940s.

The impact of insecticide use on the quality and adequacy of the food supply, while great, has not been definitely quantitated. Annual worldwide crop losses due to insects were estimated at $75 billion in 1967, or approximately 35% of production (Kulkarni and Ashoke, 1990). Borlaug (1972) estimated United States crop losses at 50% if pesticides were banned. This was disputed by a report from Cornell University (1972) estimating crop losses of about 7% if pesticides were banned. In 1972 dollars the Cornell study meant $2.1 billion losses; in 1993 dollars this would be much larger. The difficulty in estimating the real impact of removing pesticides from agricultural use is uncertainty about the portion of target crops treated, and of the extent of the damage that would be caused by insects or by variables such as weather and environmental factors beyond the control of the researcher. Estimates encompassing this number of variables tend to be very imprecise.

While pesticide usage has not solved all the problems caused by insects, or cut crop losses to zero, the fact remains that pesticide usage has, in the short term, dramatically reduced the effects of insects on crops. On the other hand, the use of pesticides in agriculture and public health to control vectors for malaria in India has not correlated well with the incidence of this disease in man. During the mid-1970s, despite greatly increased pesticide usage, the incidence of malaria increased markedly (Mehrotra, 1986).

While debate about the efficacy of pesticides continues, consumers concerned about the safety of their food supply are being heard. The 1988 Food Marketing Institute poll showed that 75% of consumers were concerned about pesticides in food, a higher percentage than were concerned about fats, cholesterol, salt and additives (Kulkarni and Ashoke, 1990).

Jurisdictions in both Europe and North America are not only considering safer new pesticides, they are also developing alternate strategies for food production which reduce or eliminate pesticides. While the subject of alternative means of controlling insects in livestock is beyond the scope of this review, the potential impact of these practices on the frequency of residue problems in livestock products must be recognized. Methods being tested include breeding of resistant livestock, release of infertile males into the environment to depress the rate of insect reproduction, hormone bioanalogues, hormone antagonists, pheromones, and changing methods of husbandry (Gunn, 1976). Recently, partial success was achieved using immunization therapy for the treatment of the bovine warble fly, *Hypoderma bovis* and *lineatum* (Baron, 1986), *Stomoxys calcitrans* (Webster et al., 1992) and *Ostertagtia ostertagi* (Christensen et al., 1992). These three parasites alone represent many millions of dollars in pesticides applied on and around livestock.

Herbacides are also pesticides; they are usually chemically and toxicologically quite different from insecticides and are not included in this review. Livestock do not usually have direct contact with these agents, so exposure is rare. The report of Smith and Lewis (1988) on incidents of pesticide poisoning in livestock didn't list any associated with herbicides and the annual report of the National Animal Poison Control Center (1986) indicated that poisoning with insecticides was more common than with herbicides in cattle and hogs. While changes in pesticide usage have had a long-term influence on the incidence of residues in food of animal origin, banning chlorinated hydrocarbons in North America resulted in a gradual decline in the frequency of residues of these agents in food over the past two decades (Frank et al., 1985).

10.3 CHARACTERISTICS OF THE MAJOR INSECTICIDE GROUPS

The insecticide classes which replaced the chlorinated hydrocarbons for use on and around livestock in Canada were the pyrethrins, organophosphates and carbamates. In Ontario these three groups make up almost all the insecticides available (Table 1). Only one organochlorine, lindane, remains on the list. A similar range of chemicals is available in most developed countries of the world. As indicated above, organochlorines remain in use for controlling *Anopheles* and *Culex* spp. mosquitoes, which are vectors for malaria in some tropical regions.

The organochlorine insecticides have been divided into three groups (O'Brien, 1967) – DDT and analogues, cyclodienes (e.g., heptachlor and methoxychlor) and a miscellaneous group which includes hexachlorohexane (a benzene derivative) and toxaphene (a polychloroterpene). As a group, they are not highly toxic to livestock. Their most serious drawback is slow breakdown in living biological systems, or when free in the soil, water or air. This slow disappearance, plus their tendency to accumulate in animal or plant lipids, causes them to biomagnify in the food chain and in the environment (Menzer, 1991). Also, some have been reported to be potential carcinogens, such as heptachlor, chlordane (Epstein, 1976), toxaphene (NCI, 1979), DDT and PCB (Falck et al., 1992). Although these chemicals have been available for many years, their precise mechanisms of action are not fully understood. Their general effect on insects is to increase the excitability of muscles and nerves leading eventually to paralysis and death (Narahashi, 1986). The mechanism of acute toxicity appears to be similar in the mammalian host (Narahashi, 1986).

The organophosphates tend to be much more acutely toxic as a group but are not very stable in either living biological systems or when free in the environment. They do not biomagnify in the food chain. They can be divided chemically into two groups—the active oxygen analogues with a $P=O$ group at the center of the chemical (e.g., dichlorovos and paraoxon) and the poorly active thiophosphates with a thiophosphate group ($P=S$) at the center of the molecule. This group must be converted to $P=O$ by oxidative phosphorylation in order to become active (e.g., parathion and malathion) (Taylor, 1990; Ecobichon, 1991). The mechanism of acute toxicity of organophosphates is irreversible inhibition of the cholinesterase enzymes (Holmstedt, 1963) responsible for the normal breakdown of free acetylcholine (ACh) at the cholinergic nerve terminals. Acute toxicity, caused by excessive acetylcholine at synapses in muscarinic nerves, is evident as parasympathetic nervous system hyperactivity (Taylor, 1990; Ecobichon, 1991). Binding to ACh nicotinic receptor sites, or at sites associated with ionic channels in close proximity, has also been reported and is associated with enhanced receptor desensitization, open channel blockade and, on occasion, agonist actions (Narahashi, 1985). Treatment of organophosphate poisoning involves the use of antimuscarinic drugs, usually atropine, and cholinesterase reactivators such as 2–PAM (pralidoxime) plus support of essential life systems (Taylor, 1990; Ecobichon, 1991).

The mechanism of action and signs of acute toxicity caused by carbamate insecticides are generally similar to the organophosphate insecticides, except that the cholinesterase enzymes are reversibly bound. The reversible nature of carbamate-acetylcholinesterase binding means it is more amenable to drug antidotal therapy, and chances of recovery are enhanced. Treatment of carbamate insecticide poisoning usually involves antidotal therapy with atropine as well as support for essential life systems (Ecobichon, 1991).

The last major group is the pyrethrin family (botanical insecticides).

They were originally obtained from flowers of the chrysanthemum family and are the oldest insecticides known to man. They are rapid knockdown insecticides and are relatively nontoxic to mammals and birds. As well, they do not bioaccumulate in the environment and are not carcinogens (Ecobichon, 1991). Recently, a group of synthetic pyrethroid analogues have been produced. The pyrethrin I analogues have no cyano group at the alpha position. For example, allethrin and tetramethrin and the pyrethrin II analogues contain a cyano group – fenvalerate and cyphenothrin (Narahashi, 1986). The mechanism of action of the pyrethrins is not well understood but appears to be similar to the chlorinated hydrocardon DDT in that they alter Na current kinetics (Narahashi, 1985).

10.4 USAGE OF INSECTICIDES IN LIVESTOCK

Livestock raised on pasture in North America and other parts of the world are subject to predation by many flying insects which can do serious damage and cause production losses. For example, the face fly *(Musca autumnalis)* and horn fly *(Haematobia irritans* or *Lyperosia irritans)* cause major problems in cattle both on pasture and, on occasion, in confinement yards. The impact of these insects includes the stress on the animals as well as the spread of diseases such as infectious keratoconjunctivitis – pink eye – caused by *Moraxella bovis.* Reduced weight gains and milk production are also problems, albeit less consistently documented. Loss of productivity depends very much on the extent of the insect infestation, and minor insect problems cannot be identified as a source of production loss (Grimson et al., 1987; Campbell et al., 1987). The warble fly *(Hypoderma bovis* and *Hypoderma lineatum)* against which many insecticides have been used (Grimson et al., 1987; Khan and Koruzub, 1981) causes significant economic losses in cattle on pasture in many areas of the world. Prior to the 1978 Warble Fly Order losses to the cattle industry in Britain ran into many millions of pounds due to damage to 50–60% of the cattle hides (Tarry, 1981). Occasional death losses have also been reported when warble fly larvae migrate through vital organs such as the spinal cord (Belle and Laval, 1981). Use of warbicides was shown by Khan and Kozub (1981) to significantly increase weight gain in feedlot cattle. In warmer climates, *Boophilus microplus* is a serious problem in cattle (Leibisch, 1986; Hopkins et al., 1985).

Treatments to control face and horn flies usually employ medicated ear tags (Hall et al., 1983; Hogsette and Ruff, 1986), aerosols (Easton, 1987; Kerbabaev et al., 1987) and/or area treatment (Skoda et al., 1987). For *Boophilus* control, pour-on synthetic pyrethroid flumethrin has been effective (Leibisch, 1986; Hopkins, 1985). Numerous insecticides have been used against the *Hypoderma* spp. (Grimson et al., 1987; Khan and Koru-zub, 1981) with the more recent treatments being directed primarily at obtaining systemic levels of the pesticide to kill the migrating form of the parasite.

Bowen et al. (1982) surveyed pesticide use in sheep dips to control insects—mainly fall head fly—in Scotland. They calculated that in 1978 42 tonnes of organophosphates and 17 tonnes of organochlorines were included in dips.

Insects which are not free-flying, such as mange and lice, are also common on all classes of livestock. Several species of mange mite are encountered in cattle and sheep *(Sarcoptes scabiei, Demodex, Psoroptes ovis* and *Chorioptes bovis)* and in swine, where *Sarcoptes scabiei* var *suis* is common. Infestations of mange mites are normally encountered in animals raised in confinement, and treatment usually involves topical insecticide application (Hewett and Heard, 1982; Curtis, 1985; Wright, 1986). More recently topical (Barth and Preston, 1985) or injection application of the anthelmintic ivermectin (Meleney at al., 1982; Ibrahim and Abu-Samra, 1987) has proven effective. Because of the inability of mange mites to fly, transmission between animals is not a problem when on pasture at low stocking densities. However, outbreaks occur when they are confined in barns or are maintained in crowded yards. There is an ongoing problem with mites in poultry production, especially layer flocks, where mites attack birds and cause feather loss, skin irritation and reduced egg production. In Canada and the United States the northern fowl mite *(Ornithonyssus sylviarum)* attacks birds in cages (Hall et al., 1983), in coops on range and in deep-litter housing. Treatment usually involves spraying the birds and their surroundings, i.e., cages or roosts, with insecticides such as pyrethrins (Hall et al., 1983; Arthur and Axtell, 1983).

Lice that live on cattle, such as blood sucking lice *(Haematopinus, Linognathus* and *Solenoptes* spp.) and biting lice *(Damalinia* spp.), on sheep *(Linognathus* spp.*)*, on swine *(Haematopinus suis)* and on chickens *(Lipeurus caponis* and *Lipeurus heterographus)* are all capable of causing disease or decreased production when infestations reach critical levels. Usual treatments include spraying, washing or contact with back-rubbers using agents such as propetamphos (Bramley and Henderson, 1984), lindane, diazinon and amitraz (Curtis, 1985). Invermectin, mentioned above for the treatment of mange, can also kill lice. It is usually injected for this purpose (Barth and Preston, 1985).

Sheep are also subject to "ked" and tick infections (Bramley and Henderson, 1984). Usually, treatments include dipping in an insecticide solution such as lindane or propetamphos (Bramley and Henderson, 1984).

As well as the problems of insects directly parasitizing livestock, the farmer must also deal with stable and house flies *(Stomoxys calcitrans* and

Musca domestica) which cause serious problems for livestock and for people working with the animals. Cattle can become difficult to work with because of the anxiety and stress caused by the flies. Milk letdown can be inhibited and flies can spread disease within barns and between farms (Quisenberry and Foster, 1984). It appears that more insecticides are used to control flies in and around livestock barns than are used to kill insects directly parasitizing the animals. The list of fly control products include pyrethrins, piperonyl butoxide, methomyl, dimethoate, dichlorovos, malathion and carbaryl (Table 1). In addition to those available in Canada, cypromazine, dimethoate, fenthion, dimethoate, permethrin and stirofos (Quisenberry and Foster, 1984) and etafos (Kerbabaev et al., 1987) are available in some other countries.

10.5 CAUSES OF RESIDUES

For chemicals to become residues in animal products they must be taken in by the animal more rapidly than they are removed. Also, the animal product must be made available for human consumption before the chemical has had time to leave the animal's system or before the chemical has been destroyed in the product during processing. The absorption of organic acids and bases, which includes most pesticides, is influenced by lipid solubility, pH of the absorbing environment, dissolution of the formulation at the absorbing surface, dispersion in the absorbing medium, and the presence of binding agents in the bowel or on the skin which can prevent uptake into the body (Baggot, 1977). Lipid solubility is the property of the chemical, while pH, dispersion and binding agents are usually characteristics of the site of absorption. Dissolution and disintegration are properties of the pesticide formulation to which the animal is exposed as well as the site of application (Benet and Scheinir, 1985). Substances such as mineral oil, activated charcoal, divalent cations, phytates and phytites are all known to decrease the absorption of drugs and chemicals in animals following oral exposure (Goldstein et al., 1974). Host factors, such as transit time in the bowel for oral exposure, or time on the skin for topical exposure, can also influence uptake.

Biotransfer and *bioaccumulation* are terms used to express the likelihood of a chemical such as a pesticide becoming a residue in an organism, be it a garden vegetable, an animal used for human consumption or man himself. Biotransfer and bioaccumulation are dependent on a chemical's stability in the environment, lipid solubility, absorbability and stability in living biological systems. Storage of highly lipid-soluble substances usually occurs in the fatty tissues of the animal's body, and in some cases, such as organochlorines, they can remain sequestered for long periods.

Bioconcentration factors are calculated as the ratio of the concentration of residue in the animal product to the concentration of the chemical in the animal's diet. Biotransfer factors are constants calculated as the ratio of the concentration of chemical in the animal product to the animal's daily intake (in mg/day/kg), determined as the sum total of residues from all sources—food, air and water. Biotransfer constants infer a rather constant daily intake of a chemical (Travis et al., 1987). Fries et al. (1973) calculated a Bmf (biotransfer constant for milk fat) for PCB at an intake rate of 0.3 and Bm for milk (assuming 4% butterfat in milk) of 0.012. The Bbf (bioconcentration for beef fat) for PCB is 0.61 and Bb is 0.15. Lipid solubility of a chemical can be expressed by the octanol/water partition coefficient (Kow) and a direct correlation has been calculated between Kow and the bioconcentration (B) of chemicals in animal products. The linear relationship between the Kow and bioconcentration of a chemical in beef (Bb) is log Bb = −7.679 + 0.996 log Kow. If the slope is constrained to 1, log Bb can be described as −7.6 + log Kow as calculated by Travis et al. (1987). Similar biotransfer equations have been calculated by these authors predicting residue transfer to milk based on Kow (log Bm = −8.144 + 0.992 log Kow).

Besides the factors mentioned above which promote or are essential for the production of a residue in a food product of animal origin, there are a number of situations which involve an unusual use of a chemical, or represent loss of control of a chemical in the environment, that can lead to a residue in a food. These may include one or more of the following:

(1) The failure of the producer to follow appropriate withdrawal periods even though appropriate guidelines are available
(2) The application of pesticides to the wrong class of livestock
(3) The presence of pesticide residue in the animal's feed or water as a result of contamination of the environment or random events, e.g., spills of chemicals on feed causing high level exposure of the use of unconventional feeds
(4) Inadequate or incorrect withdrawal data

The fourth possibility seems very unlikely given the present level of research and technical information on pesticides. One and two are the most likely causes of the residues detected in surveillance programs (USFDA FSIS Science Program, 1989). Problems of widespread environmental contamination with organochlorines remain in many developing countries, and these cause unusually high background levels of the pesticides in human food products (Dhaliwal, 1990). Also, improvements in sampling and detection technology have turned up unexpected residues, or residues at concentrations that were previously undetected.

Exposure through the diet is of particular concern, since ruminants may pasture or drink surface water in areas contaminated by pesticides used to control weeds or insects. They may also consume water or forage contaminated by industrial effluent. Animals fed apple pomace, pineapple greenchop, pea vine silage or bakery waste, etc., may encounter pesticides for which there is no residue information. These by-products of agriculture and industry are not considered traditional animal feeds and therefore residue data is frequently not available.

A survey of the literature on the frequency of pesticide residues in human food of animal origin suggests that the incidence is quite low in some countries and quite high in others. During the early 1970s a serious problem occurred in the United States involving residues of polybrominated biphenyl (PBB) in the milk and meat of dairy cows. It was centered mainly in the Michigan area (Welborn et al., 1975; Mercer et al., 1976). Fire Master FF-1, a commercial fire retardant, was accidentally mixed in livestock feed, and before the mistake was discovered milk from contaminated cows had been dispersed widely, exposing the public to potentially harmful residues. This incident led to the destruction and burial of 29,800 head of cattle as well as lost production and milk sales costing the industry many millions of dollars. Examination of the uptake and excretion of PBB indicated that approximately 50 percent of an oral dose was excreted in the feces and a little over 20 percent in the milk. No free PBB was found excreted in the feces although a significant amount of the PBB metabolites were excreted by this route (Welborn et al., 1975).

In recent years the problems of Australian beef transported to overseas countries have been well publicized (Corrigan and Seneviratna, 1989). Residues most frequently reported were the organochlorines, DDT, heptachlor and dieldrin. These contrast with the reports of no significant violative residues in Canada (Saschenbrecker, 1976), USA (Steffey et al., 1984), West Germany (Hecht, 1985), and Italy (Maderena et al., 1980). Al-Omar et al. (1985) and Davies and Doon (1987) reported violative residues of dieldrin and heptachlor in Iraqi beef, and DDT in beef from El Salvador.

The response of the Australian government to the problem was to greatly increase surveillance and testing in slaughterhouses, and to initiate a traceback and quarantine program for affected farms (Corrigan and Seneviratna, 1989). Once the surveillance program was in place, the incidence of violative residues dropped rapidly. The occurrence of organochlorine residues in meat reflects the continued use of the agents to control insect pests, although very recently control measures have been tightened. Dickson et al. (1983) reported an episode of heptachlor poisoning in cattle and horses grazing a field on which commercial heptachlor had been applied at a rate of 6.5 kg/hectare. Renal fat concentrations in poisoned animals ranged from 180–500 mg/kg. Harradine and McDougall (1986) examined steers grazed on fields which for the previous 2 years had been treated with heptachlor just prior to planting potatoes and maize. During the 19 months that the animals were studied on the pasture, the maximum residue concentration reached in the fat was 0.72 mg/kg. These authors reported that intake of heptachlor was greatest when the pasture was short. They suggested that the main source of residue was soil consumption by the cattle, and this was increased when the grass was closely cropped. Soil dust on the foilage was also considered an important source. Petterson et al. (1988) studied an incident of beef cattle being exposed to heptachlor while grazing on land sprayed to control ants in Argentina. A commercial solution of 0.5% heptachlor was applied at a rate of 1300 L/hectare. One animal died; the others were moved to a clean paddock for a time and then to a research station for observation and monitoring. Initial levels in subcutaneous fat were heptachlor epoxide 22 mg/kg and oxychlordane 5.68 mg/kg; these dropped to 0.08 mg/kg and 0.18 mg/kg respectively over the 488-day observation period. These authors indicated that a high plane of nutrition appeared to result in lower levels in the fat, either by enhanced removal of the material from the animals' systems or by dilution in the fat. Lactation had no effect. Neumann (1988) reported that the most frequent organochlorine found in the surveillance of Australian beef was dieldrin (72.4% of all residues), next was heptachlor (11.2%) followed by DDT (7%) and small amounts of chlordane and benzene hexachloride. Tracebacks to the source revealed that 46.6% of the sources were chemicals sprayed on pasture, 32.5% were from termite control, and 10.2% were found in feed and storage areas.

Samples of perirenal fat from sheep raised in France, when tested for lindane, were below the EEC directive level of 2 mg/kg. Analysis of perirenal fat from sheep not raised in France revealed that 17% of the animals were above the accepted standards. The highest residue found in the survey was 9.5 mg/kg. Most of the positive residues were observed during August and September (Venant and Richou-Bac, 1985).

Chlorinated hydrocarbon contamination of the feed was observed on six Arizona dairy farms by Whiting et al. (1987). The source of residues was difocol, an ascaricid composed of kelthane and trace amounts of DDT. Violative residues of chlorinated hydrocarbons were not found in the milk. The main residues detected were DDE (dichloro-diphenyl-ethane), the primary metabolite of DDT, and kelthane.

Dhaliwal (1990) reviewed the status of organochlorine residues in milk and butter in India and other developing countries. Surveys throughout the 1970s and 1980s reveal almost universal contamination of the milk supply with DDT and metabolites p,p'–DDE, and p,p'–TDE. When examined on a milk fat basis, a very high percentage of the residues were above the maximum residue limit (MRL) for DDT (Kalra and Chawla, 1986). Goat and buffalo milk in India seem to be contaminated to the same extent (Saxena and Siddiqui, 1982). Widespread contamination of the milk and milk products in India must be viewed in light of the continued use of DDT in malaria control programs. Public health use of DDT in India was reported by Mehrotra to be approximately 10,000 tonnes/annum and another 2,000 tonnes/annum is used in agriculture (Mehrotra, 1985a).

Other chlorinated hydrocarbons besides DDT have been found with great frequency in India. HCH isomers were detected in virtually all milk samples collected in the Punjab and analyzed by Kapoor et al. (1980). Similar findings were obtained by Singh et al. (1986) in cow's milk and by Kaphalia et al. (1985) in buffalo milk in the Lucknow region. Aldrin (Saxena and Siddiqui, 1982) and dieldrin (Khandekar et al., 1981) levels as high as 500 times the MRL have been found. Endosulfan was found in milk in 3 of 105 samples taken from Hisar. Levels were as high as 2.5 ppm (Chauhan et al., 1982).

Cows milk produced in Japan at one time showed evidence of DDT contamination (Uyeta et al., 1970). Milk from three Central American countries, El Salvador, Guatemala and Honduras, contained residues as high as 32.3 ppm when examined by Mazariegos (1976).

As was previously discussed, the organophosphate insecticides have a reduced potential for residues in food products because they tend to be more rapidly eliminated from the animal's body, and they are not very stable in the environment (Matthees, 1990). Muan et al. (1985) examined the milk of Norwegian red cows following topical sponging with a 0.5% malathion emulsion. Maximum levels of 0.054 mg/L were found 4–6 hours after treatment. By 24 hours the concentration was 0.005 mg/L and by 72 hours after treatment the levels were 0.002 mg/L. Approximately 0.006–0.03% of the dose appeared in the milk. A single application of coumaphos dusting powder, at a dose of 50 grams of the 1% powder per cow, gave higher residue levels. Nijhuis et al. (1984) indicated that phoxim, 500 mg in 3000 mL of solution, applied topically to treat scabies in cattle, caused residues in milk of 0.02 mg/L at the first milking. Since 1 L of this milk did exceed the acceptable daily intake (ADI) for a 70 kg man—0.014 mg—a withholding period was required after treatment.

The annual report of the National Animal Poison Control Center (1986) indicated that insecticide residues were found in approximately 10% of the reports or inquiries of bovine insecticide poisoning. The most common residues were heptachlor, 8 out of 18, famphur, 2 out of 6 and fonfos, 1 out of 6. In swine, residues reported were methoxychlor, 2 out of 5 cases. Captan, an agriculture fungicide, was found in 2 out of 11 bovine cases and 2 out of 3 swine cases.

When listing situations where residues are likely to become a problem in livestock products, the use of nontraditional feeds for which there is no information regarding withholding times was mentioned. Ruminants, for example, may be fed materials such as apple pomace, pineapple greenchop or pea vine silage. These feeds can be contaminated with pesticides used during their cultivation. Frequently, there are no published withholding times for these types of materials. The impact of potential use of pesticides on these type crops to consuming animals needs to be assessed. Chase et al. (1987) showed that dicofol, used in apple orchards, was capable of causing residues of the parent material (kelthane), plus traces of *pp'*–DDE, in milk of cows consuming apple pomace. The source of *pp'*–DDE, the prime metabolite of DDT, is a small amount of DDT present (2.5 percent) in the commercial dicofol formulation. Fox et al. (1989) observed that some commercial dairy herds fed apple pomace had detectable levels of kelthane and *pp'*–DDE in the milk as well as in the fat of slaughtered animals. Maximum levels observed in fat were 61 ppb *p,p'*–DDE and in excess of 250 ppb kelthane. None of the residues represented a health hazard to the consumer.

10.6 REGIONAL DIFFERENCES IN PESTICIDE RESIDUES

Dogheim et al. (1988) reported on a survey of pesticide residues in milk and fish collected in upper Egypt. The Beni-Seuf governate region is a mainly agricultural area with extensive use of pesticides on crops such as cotton. There is little or no contamination of the environment from other industries. Buffalo milk contained relatively high mean levels of organochlorines, i.e., total hexachlorohexane isomers 8.574 ppm, total DDT isomers 3.403 ppm, total heptachlor isomers 0.155 and aldrin + dieldrin 0.521 ppm. No PCB's were found in the study. Catfish and boltifish were collected from the local markets since they are an important food source for farmers. Residues were less evident in the fish than in buffalo milk and the levels were generally lower. DDT metabolites made up a much larger portion of the total DDT in fish than in the buffalo milk. The ratio of DDT (*o,p'* and *p,p'* isomers) to metabolites (*p,p'*–DDE and *p,p'*–DDD) was 3.029/1 indicative of relatively recent exposure. By contrast, in Canada where DDT has not been used for many years, only trace amounts of the parent compound are detected in milk and meat and the residue is mainly metabolites (Frank et al., 1975, 1983, 1985a, 1985b).

Frank et al. (1985a) surveyed pesticide residues in milk from Ontario farms. Total DDT residues on a county basis ranged from a high of 0.033 ppm in Essex to a low of 0.00366 ppm in milk from Thunder Bay. The Essex county area is very intensively farmed, with a great deal of cash crop production. The Thunder Bay area in northern Ontario has rather sparse agriculture and much less cultivated field crop production. The metabolite of DDT, p,p'–DDE, made up at least half of the total DDT reported and the ratio of DDT to metabolites was less than one. This indicates that the ban on the use of this compound has had its effect. The DDT causing the residues had been in the environment for a long time and was gradually broken down.

It is of interest to note that the data from upper Egypt indicated the absence of PCB in any products sampled. This area, as noted above, is industrially underdeveloped (Dogheim et al., 1988). In contrast, the Niagara area of southern Ontario had 0.130 ppm PCB in milk fat, well in excess of any other area of the province. The Niagara area is known for its intensive fruit and vegetable production, and is located between the heavily industrialized areas of Toronto and Hamilton, Ontario and Buffalo, New York. It is home to some of the major hydrogenerating stations in North America. While the source of the PCB has never been located, the proximity to the industrial and electrical generating activity, as well as to major toxic dump sites in New York state, has been noted.

10.7 INTERNATIONAL RESPONSE

International interest in pesticide use in agriculture has been around for many years; the FAO pesticide program was initiated in 1959 (FAO, 1959). At a joint meeting of the World Health Organization (WHO) Expert Committee on Pesticide Residues and the Food and Agriculture Organization (FAO) Panel of Experts on the Use of Pesticides in Agriculture held in 1961, it was recommended that studies be undertaken to evaluate possible hazards to man arising from the occurrence of pesticide residues in food. The Joint Meeting of the FAO Committee on Pesticides in Agriculture and the WHO Expert Committee on Pesticide Residues (JMPR) occurred again in 1963, the mandate being to evaluate hazards to consumers resulting from pesticide residues in food. The JMPR has met regularly since its inception to advise others on the potential hazards of pesticides. It is composed of experts drawn from member countries of FAO and WHO acting as private individuals. Pesticides are evaluated on the basis of all available data, supplied mainly by governments and industry, and, where appropriate, human ADI's (acceptable daily intakes) for the chemicals and MRL's (maximum residue limits) for foods are established. In the 1963 and 1965 JMPR meetings suggestions for ADI were made for 14 different pesticides. These were chemicals used mainly in production and/or storage of cereal grains such as DDT, lindane, carbaryl, malathion, parathion, and diazinon (FAO, 1966).

At approximately the same time, 1963, a FAO Working Party on Pesticide Residues was established with the purpose of advising the director general on all matters related to pesticide residues with particular reference to toxicity of pesticides, analysis of pesticide residues, pesticide tolerances and surveys for collecting residue data (FAO, 1966).

One role of the JMPR has been to recommend ADI's and MRL's to the Codex Committee on Pesticide Residues (CCPR). The CCPR is responsible for setting the MRL's which are, with the approval of the Codex Alimentarius Commission, published. The guidelines for MRL's assume that the residues which appear in the food will be those resulting from the use of the chemical on a global basis according to good agricultural practice (GAP). *GAP* means the adherence to recommended or authorized use of the pesticide in food production, be it the animal husbandry, cereal production, transportation or storage of food, or in the food inputs. As in the case of ADI's, MRL's are subject to change as new scientific information comes available, or after experience is obtained with the chemical in the field.

The Codex Alimentarius Commission first met in 1963. It was established under the auspices of the United Nations Environmental Protection Agency (UNEP), the FAO and the WHO to implement the Joint FAO/WHO Food Standards Programme (Codex Alimentarius Commission, 1973). Thirty countries joined the commission at its inception; it has since expanded, and the 1981 membership was 129 countries (Codex Alimentarius Commission, 1985). A number of Codex committees have been struck to advise the commission on topics of special interest. The Codex Committee on Pesticide Residues (CCPR) was established to advise the Commission on:

(1) MRL's for pesticides in the various classes of foods
(2) Priority lists of pesticides
(3) Sampling and analysis techniques
(4) Other matters relating to safety of pesticides

Recommendations of the JMPR are used by the CCPR as bases for recommendations concerning safety of pesticide residues in food (Codex Alimentarius Commission, 1978).

10.8 ADI's and MRL's

The ADI is the daily dose of a chemical that, over an entire lifetime of exposure, is "without appreciable risk" to the consumer—on the basis of all factors known at the time. "Without appreciable risk" has been taken to mean the practical certainty that injury will not result after a lifetime of exposure. It is calculated on the basis of available toxicological data from animal and human studies as well as observations of the effects on man. Starting point is the calculated no-observed-adverse-effect-level (NOAEL) of the most sensitive toxicological parameter in the most sensitive species. To derive an ADI, a safety factor is then applied to this value to account for interspecies variation, statistical, experimental and other factors. Conditional ADI's are given when uncertainty about the effects exist, usually as a result of inadequate experimental data. Monographs are prepared on pesticides and they are updated with new information as warranted.

MRL's under consideration or recommended by the Codex Alimentarius Commission are published in the Guide to Codex Maximum Limits for Pesticide Residues. The first issue was 1978 (Codex Alimentarius Commission, 1978). In it food commodities are grouped and coded and, if available, MRL's listed for the food group. As an example, the coding for poultry meat B.07.2500 is class (B) for animal, (07) for poultry and (2500) for poultry meat. MRL's cover pesticides likely to be a problem in the target commodity (Codex Alimentarius Commission, 1978).

The fact that a country is a member of the Codex Alimentarius Commission does not mean that it accepts all published MRL's as standards for commerce, either within its borders or with other countries. This is readily apparent from the minutes of the numerous general meetings (Codex Alimentarius Commission, 1985). The trend, however, in international commerce is clearly to greater and greater acceptance of these guidelines.

10.9 RESIDUES AND FOOD INTAKE DETERMINATIONS

In order to assess the likelihood of residents of an area receiving greater than the ADI of a residue, it is essential to know the daily food intake and to relate it to total residue intake in food. Numerous difficulties have been encountered when attempting to calculate daily intake of pesticide residues. Uncertainty about food intake and rate of exposure translates into uncertainty about residue exposure. Difficulties exist not only between countries but also between geographical areas in the same country, between socioeconomic levels and indeed within family units. Not surprisingly, gender is a factor. Males eat more meat than females from the end of the first decade of life until death. Vegetable intake is approximately the same but females consume larger quantities of fruit. Age is also a consideration; infants consume more dairy products than adults (Salathe and Buse, 1979). Differences in body weights and food availability further magnify the problem. Countries are urged to calculate their own daily intake values on the basis of cultural diets rather than utilize figures from other regions of the world (UNEP/FAO/WHO, 1989).

The potential for a food residue to cause harm can be assessed by comparing the ADI and the estimated daily intake (EDI). The MRL for a chemical is the maximum residue in a crop or food product used according to GAP. It does not, at least superficially, include information on dietary intake and toxicity. Authorities from time to time are called upon to estimate intake of certain chemicals by their population and to compare them with ADI's. Four options have been proposed for calculating the daily intake of pesticide:

(1) The most accurate is the direct daily measurement of intake.
(2) The second-most accurate is the *best estimate* of the ESI.
(3) The third-most accurate is the *intermediate estimate* called the estimated maximum daily intake (EMDI).
(4) The least accurate estimate is the *crude estimate,* the theoretical maximum daily intake (TMDI) (UNEP/FAO/WHO, 1989).

The EDI calculation includes the residue levels detected in foods, the proportion of the crops treated—husks, stalks, etc.—and the amount of purchased versus homegrown food that is consumed. The effects of cooking, storage and processing on residue levels are also included in the calculation. Food consumption is calculated for the subgroup being evaluated. Clearly the calculation of EDI requires a great deal of information. In many instances it is not available.

The EMDI is calculated by the following realtionship: $EMDI = "(F_iR_i-P_iC_i)"$ where F_i = amount of each food consumed in kg/person/day, R_i = the residue concentration mg/kg in the proportion of the food eaten, P_i is a correction factor for loss of residue in processing and C_i is a correction factor for the loss of residue in cooking. Correction factors are the given in the JMPR evaluations. "E"n indicates that EMDI is the sum of the residues in each of "n" foods consumed in a day. The theoretical maximum daily intake of a pesticide can be calculated using the formula $TMDI = "E"n \ F_iM_i$. TMDI is the sum of residues in "n" foods consumed on a daily basis. The amount of the commodity is based on global or national food consumption data (in kg/person) and it is multiplied by the M_i which is the MRL for the food. The units of MRL are mg/kg (UNEP/FAO/WHO, 1989).

Since the most accurate approach of actually measuring the residue levels in the diet is impossible on a broad scale, countries are forced to estimate the residue intake of its citizens using one of the above methods. The EDI calculation is valid only for the target population and must be interpreted in that manner. It should not be extrapolated to populations which have different cultural diets or eat food from greatly differing sources. EMDI tends to overestimate the total residue in the population's diet because it assumes that all food eaten is treated with a pesticide and that all food contains the MRL level of pesticide. Usually only a small proportion of a food supply is exposed to a given pesticide, and of the food exposed only a very small proportion will have concentrations equal to the MRL. Thus, the EMDI is very conservative and tends to overestimate the actual pesticide burden of the diet. The TMDI calculation tends to overestimate the dietary pesticide intake the most. In contrast to the EMDI it makes no allowance for loss of pesticide in cooking or processing, and assumes that the food eaten contains residue at the same level as the whole material. Most food eaten does not contain the outer covering, i.e., skin, shell or husks where residues are normally the highest. In addition to these limitations, the problems of using the MRL as the residue level and assuming that all the crop has been treated with pesticide also exist in the TMDI calculation.

Examination of recent reports of the Food Safety and Inspection Service (FSIS) in the United States indicates that the incidence of residues is very low and indeed few instances of MRL or greater are detected (USFDA FSIS Science Program, 1989). Most have no residue detectable. Dietary intake calculations assuming that food contains MRL's far overestimate the chemical intake and err on the side of safety. Comparison with the ADI and the EDI must be on the basis of average daily intake rather than exceptional circumstances where the intake is very high since the ADI estimation reflects a lifetime of exposure.

10.10 NATIONAL RESPONSE

In the United States, the Food and Drug Administration (FDA) and the United States Department of Agriculture (USDA) have the major responsibility for insuring that the food supply is safe and free from pesticide residues. The Environmental Protection Agency (EPA) acting under the Food, Drug and Cosmetic Act (FDCA) is responsible for the registration of pesticides under the Federal Insecticide, Fungicide and Rodenticide Act (FIFRA). Required information includes chemical composition, formulation, manufacturing process, analytical methods, plans for usage and safety data. Safety data includes information on carcinogenicity as required by the Delaney Amendment which prohibits additives in food that induce cancer in humans or animals (Committee on Scientific and Regulatory Issues Underlying Pesticide Use Patterns and Agricultural Innovation, 1987b). Data from toxicity and residue studies is also required under the Food, Drug and Cosmetic Act. The least dosage to produce an effect in the most sensitive parameter is used to establish a no-effect level (NOEL). A safety factor of from 10-10,000 is applied to the NOEL to calculate an acceptable daily intake. The safety factor takes into account the fact that the data are derived from animal experimentation rather than human studies (Kulkarni and Ashoke, 1990). The monitoring of the meat and poultry for pesticide residues is a function of the FSIS under the Federal Meat Inspection Act (FMIA) and the Poultry Products Inspection Act (PPIA) of the USDA (USFDA FSIS Science Program, 1989).

The responsibility for control of pesticide residues in Canada appears to be less focused than in the United States. The Pesticide Control Act which registers pesticides in Canada is under the control of Agriculture Canada (Frank and Ripley, 1990). Data on the safety, efficacy, environmental impact, residues, residue detection methods, carcinogenicity, etc., are all required under this act before registration can occur. Agriculture Canada also controls the use of pesticides and pesticide residues in the various food commodities under the Feeds Act, the Seeds Act and the Meat Inspection Act. The safety of pesticides is determined and ADI's and MRL's developed by safety evaluators in Health and Welfare Canada (Health and Welfare Canada, 1988).

The potential for a product to affect the environment is assessed by Environment Canada and the impact on fish and aquatic organisms is assessed by Fisheries and Oceans Canada. Monitoring activities for residues in both imported and domestic foods are carried out by Health and Welfare Canada under the Food and Drug Act, for animal feeds under the Feeds Act, and for grains used for consumption and export under the Canadian Grain Commission. Livestock and poultry meats residue inspection is under the Federal Meat Inspection Act of Agriculture Canada (Frank and Ripley, 1990). Several provincial governments have food safety and testing programs of their own, and these provide additional surveillance of the food supply for pesticide residues (Frank et al., 1985).

10.11 RESIDUE TESTING PROGRAMS

Using ADI's and MRL's as either criteria or guidelines, plus unique national criteria, a number of countries have mounted effective monitoring programs for pesticide residues. The criteria for focus on a chemical may vary between countries, but the USDA uses the following criteria in determining the likelihood of harmful residues appearing in the food supply:

(1) Amount of chemical actually or probably used
(2) Conditions of use as related to residues at slaughter
(3) Potential for misuse to result in harmful residues
(4) Metabolic patterns of the chemical in animals, plants or the environment, including the bioavailability and persistence of residues
(5) Toxicity of the residues (USDA FSIS, 1989)

The action levels and MRL's for animal food products in the U.S. and Canada are shown in Tables 2 and 3.

Until recently the Japanese government had no residue limits for pesticides in meat products. They were forced to adopt temporary measures for monitoring beef when it became evident that Australia was having a problem with chlorinated hydrocarbon residues in some of its beef exports (Tonogai et al., 1989). The MRL's adopted were the standards already in use for vegetable products: DDT (DDE + DDD) 0.5 ppm, dieldrin (aldrin) 0.2 ppm and heptachlor (+ epoxide) 0.2 ppm. These concentrations are lower than the action levels currently in place for beef in either the United States or Canada, but are in line with MRL's for vegetable products in these countries.

The international standards for fish products differ somewhat from livestock products (Table 4). The Codex Alimentarius Commission, however, gives fish an animal product designation in group B (Codex Alimentarius Commission, 1978).

The FSIS in the United States is responsible for monitoring meat and poultry for residues of pesticides, antibiotics, metalotoxines or any chemi-

TABLE 10.2. Action Levels and Residue Limits for Organochlorine Pesticides in Meat Products in the U.S.A.

	Food Product	
Pestice	Animal* Fat	Muscle
Aldrin + Dieldrin	0.3a**	
BHC	0.3a	
Chlordane	0.3a	
DDT + DDE	5.0a	
Endrin	0.3a	
HCB	0.5a	
Hepachlor + Chlordane and Heptachlor metabolites	0.3a	0.0
Lindane	4art or 7‡	
PCB	3.0ar*§	
PBB	0.3	0.05
Toxaphene	7.0ra§§	

* Cattle, sheep/goats, swine, poultry, horses
** a = action level.
† ar = action level for poultry, residue limit swine.
‡ Residue limit cattle, sheep/goats, horses.
§ ar* = action level for cattle, sheep/goats, swine, horses and residue limit for poultry.
§§ ra = residue limit cattle, sheep/goats, swine and horses and action level poultry.
Unless otherwise indicated the numbers quoted are residue limits.
All residues in dairy products are on a fat basis.
Source: USFDA FSIS, 1989.

TABLE 10.3. Maximum Residue Limits* for Pesticides
in Food Products (ppm) in Canada.

Pesticide	Milk	Dairy Products	Butter	Animal Fat	Meat†	Eggs
		Animal Food Product				
Alachlor + metabolites	0.001				0.001	
Aldrin (f)** Dieldrin	0.1	0.1	0.1	0.2	0.2	
BHC (f)	0.1	0.1	0.1	0.1	0.1	
Chlordane (f)	0.1	0.1	0.1	0.1	0.1	
Chlorpyrophos (f)			1.0	1.0‡	1.0	
Coumaphos (f)					0.5	
Crufomate					1.0§	
DDT (f)	1.0	1.0	1.0	1.0	1.0	
Dicofl (f)	0.1	0.1	0.1	0.1	0.1	
Dioxathion					1.0	
Endosulfan (f)	0.1	0.1	0.1	0.1	0.1	
Endrin (f)	0.02	0.02	0.02			
Ethion (f)	2.5	2.5	2.5	2.5	2.5	2.5
Ethoxyquin					0.5	0.5
Flucythrinate (f)		0.1				
Heptachlor (f)	0.1	0.1	0.1	0.2	0.2	
Lindane	0.2	0.2	0.2	2.0	0.7–2.0§	
Methoxychlor (f)					3.0§§	
Tetrachlorvinphos (f)				0.75–1.5‖	0.75–1.5‖	

*The sale of foods that contain residues in excess of Maximum Residue Limits may be illegal.
**(f) calculation based on fat content.
†Meat and meat by-products from cattle, goats, sheep, hogs, poultry.
‡1.0 ppm in fat, liver and kidney in cattle only.
§0.7 ppm poultry meat and 2.0 ppm cattle, goats, sheep, hogs residues in fat content.
§§Cattle, hogs, sheep.
‖1.5 ppm cattle and hogs and 0.75 ppm poultry.
Source: Agriculture Canada, Agri-Food Safety Division, 1989–1990.

TABLE 10.4. Action Levels for Pesticide Residues in Fish.

	Concentration Range (ppm)
Aldrin	0.2–0.5
Dieldrin	
DDT and + metabolites	2.0–5.0
HCH isomers	0.2–0.5
Hexachlorobenzene	0.2–0.5
Lindane	0.5
Oxychlordane	0.01–0.03
PCB's	2.0–5.0

Reported action levels for Canada, France, Denmark, Sweden, United States and Thailand.
Source: FAO, 1983.

cal which is potentially harmful to consumers. The livestock classes monitored in the surveillance program in 1987 (Table 5) were horses, bulls, steers, cows, heifers, calves, mature sheep, lambs, goats, market hogs, boars and stags, sows, young chickens, mature chickens, young turkeys, mature turkeys, ducks, geese and rabbits. The range of residues monitored in a class depends on the probability of their presence, i.e., ducks were tested for pesticides (16 different chemicals in total), rabbits were tested for antibiotics and young turkeys were tested for antibiotics, arsenic, trace elements, sulfonamides and other compounds (USFDA FSIS, 1988). Only 14 violative pesticide residues, out of 4,326 samples, were reported in the National Residue Monitoring Program in 1987, one each of BHC, DDT and HCB; two each of heptachlor, dieldrin and nonachlor; and 5 chlordan (USFDA FSIS, 1988). In the surveillance program only 7 violative residues were observed, one DDT (metabolites) in a bull and a mature chicken and 5 heptachlor (plus metabolites) in mature chickens (USFDA FSIS, 1988).

Data on residues in livestock obtained by the USDA Surveillance Program is of less statistical relevance to the total food supply since it contains a built-in bias. Carcasses sampled in this program have been specially selected by food inspectors for investigation. Usually the carcass warrants

TABLE 10.5. Pesticide Residues: USDA Surveillance* Program (1987).

| Livestock Class | Type of Pesticide | | | |
	Chlorinated Hydrocarbon	Carbamate	Organophosphate	PCB
horses		1/0**		
bovine	1/0			
bulls	16/1			
steers	7/0			
cows	4/0		1/0	
heifers	11/0		3/0	
calves			1/0	
bob veal calves	6/0		1/0	1/0
market hogs	22/0			5/0
sows	1/0			
young chickens	13/0		1/0	1/0
mature chickens	12/6			
young turkeys				10/0
mature turkeys	1/0			
ducks			1/0	

*Samples for surveillance are taken from particular carcasses or products in response to information indicating that adulterating levels of residues may be present.
**Number of animals sampled/number of samples with violative residues.
Source: *Domestic Residue Data Book*, National Residue Program (USFDA FSIS, 1988).

special attention because of evidence of significant pathological lesions, or the situation surrounding the availability of the carcass indicates a need for testing (USFDA FSIS, 1988).

Christie (1987) reviewed the policies in Great Britain which play a role in controlling residues of veterinary drugs in food of livestock origin. The overall responsibility for this program rests with the Steering Group of Food Safety which, through its working parties, has a wide range of surveillance programs. Actual collection of samples is carried out by veterinary officers of the Ministry of Agriculture, Fisheries and Food. Samples are collected from cattle, calves, pigs, sheep, turkeys, chickens and geese. Antibacterials, hormones and pesticides are monitored in these species. In addition to this monitoring scheme, organizations such as the British Veterinary Association, the Royal Society of Veterinary Surgeons, the National Farmers Union plus other interested organizations together developed the "Code of Practice — For the Safe Use of Veterinary Medicines on Farms." The code emphasizes the importance of the proper use, proper recording of usage and compliance with prescribed withdrawal periods.

10.12 IMPACT OF RESIDUE CONTROL PROGRAMS

The overall benefits of residue control programs, restrictions on availability, residue monitoring and testing programs and restrictions on the types of pesticides available were not evident until years after their initiation. The report of Frank and Ripley (1990) demonstrated the long-term impact of restrictions on DDT and PCB use in North America on the incidence of DDT and PCB in Ontario animal products. In the 1969–70 period only 0.8% of the beef fat samples tested had DDT levels lower than 10 mcg/kg, by 1981 73.1% were below this norm and by 1985–86 all beef fat samples tested were below this concentration. Pork and poultry fat samples have mirrored this trend. Changes in PCB concentration were similar to the DDT changes. In 1969–70 no beef fat samples had PCB concentrations below 10 mcg/kg; by 1985–86 all beef, pork and poultry fat samples were below this concentration.

Pennington and Gunderson (1987) reported a similar drop in the incidence of residues in the Total Diet Study. In the 1960s chlorinated pesticides intake was much closer to the ADI than it is today. For example, dieldrin intake today is approximately one-twentieth what it was in the 1960s and early 1970s. Residues of persistent chlorinated hydrocarbon pesticides remain today. However, the levels are very low and mainly in foods of animal origin.

Corrigan and Seneviratna (1989) reviewed the problems of pesticide residues in Australian beef. In 1986 the Australian government undertook a survey of the national meat supply to evaluate the frequency of residues. They found that 0.34% had unacceptable residues (Anon., 1987). However, in 1987 there were a number of reports from off-shore countries about the presence of unacceptable levels of organochlorine residues in beef originating from Australia. A national survey followed within Australia confirming these facts, and identified the problem chemicals as DDT, dieldrin and heptaclor (Anon., 1987). The sources were mostly mixed farming enterprises in a relatively circumscribed area. Contamination of pastures was occurring through pesticide use on other crops for insect control. Animals were accumulating the residues in their body fat, over a period of years, while grazing these fields. The government response was to intensify testing of carcasses by 10-fold and to trace back and quarantine farms on which violative residues were found. Contaminated livestock were either held on the farms or moved to clean pastures until the residues reached acceptable levels. The sale and use of DDT was banned, and heptachlor and dieldrin use was limited to licensed operators for the control of termites in buildings. The incidence of unacceptable residues has fallen from 0.42% in 1986–87 to 0.22% in 1987–88 (Corrigan and Seneviratna, 1989).

10.13 IMPLICATIONS FOR HUMAN SAFETY

After approximately two decades of interest in the health effects of residues in food, we still do not know if they are of any significance to human health at the levels and frequency we are presently finding. The author was unable to find evidence of pesticide residues in food of animal origin causing health problems in humans, although there have been instances of serious contamination of food grains with pesticides causing serious poisoning outbreaks. There are two major safety issues resulting from the use of pesticides in food production or in the environment—direct toxicity and a possible increase in the incidence of cancers. Reports relating human toxicity directly to pesticides in food are rare. These occurrences usually result from the accidental contamination of food with a pesticide, or the use for human food of grain rendered unfit for human consumption, then somehow being rediverted into the human food supply. PCB-contaminated oil from a heat exchanger in a processing plant contaminated rice and caused a condition known as *Yusho* in Japan (Fries, 1972). Endrin was implicated in human poisoning when large amounts of the material contaminated flour during shipment (Hayes, 1975). An outbreak of endrin poisoning also occurred in Saudi Arabia and Quatar. Twenty-six deaths were reported; large numbers of people were affected. The consumption of seed grain treated with hexachlorobenzene by people living in south-eastern Turkey led to the development of a condition occasionally referred to as *monkey disease*. Liver damage, porphyria, cutaneous pigmentation, altered hair growth and a mortality of about 10% were recognized (Hayes, 1982).

In all documented cases where consumers have been poisoned, the contamination was directly due to human or mechanical failure in handling the pesticide in the proximity of food. The safety studies carried out by manufacturers, prior to clearance of these agents, and the government monitoring programs appear to have prevented serious toxicity problems in humans consuming foods exposed to these agents.

Despite the safety studies, an element of risk remains for workers handling, manufacturing or applying these agents in agriculture, especially when strict guidelines for safe handling are not followed. Exposure level appears to be the key element, and cancer epidemiological studies correlate positively with exposure to pesticides. For example, high levels of DDT metabolites in milk correlate with an increased incidence of breast cancer in women (Falck et al., 1993).

10.14 CANCER RISK ASSESSMENT

One of the most serious concerns of consumers is the possibility that residues of pesticides in food are causing an increase in the incidence of cancer amongst humans. This concern has been very difficult to document because of the slow development of this condition in man, the very low incidence of residues in food and the high background risk of cancer, calculated at approximately 25% over a lifetime of 70 years (National Research Council, 1984). The 1958 Delaney Clause in the U.S. Food, Drug and Cosmetic Act (FDCA) prohibits the addition of cancer-causing chemicals to food. The Environmental Protection Agency (EPA) interprets this to mean oncogenic chemicals—those that cause tumours—whereas the FDCA interprets it to mean chemicals that cause carcinoma (Committee on the Scientific and Regulatory Issues Underlying Pesticide Use Patterns and Agricultural Innovation, 1987b). Both the FDCA and the EPA, acting under the Federal Insecticide, Fungicide and Rodenticide Act (FIFA), are involved in setting tolerance levels for pesticides in food.

Similar activities and rationale are applied in many other countries of the world, although the circumstances of the application may be different.

The FDCA benchmark for judging safety is a lifetime upper-bound risk estimate of 1×10^{-6}. A chemical with this risk potential would be expected to increase the cancer incidence in one million people from approximately 250,000 to approximately 250,001. The exposure assumption involves a full lifetime of 70 years. Estimation of oncogenic risk via the diet is calculated as: dietary oncogenic risk = exposure (food intake × pesticide concentration) × oncogenic potency (potency factor or Q*) (Committee on the Scientific and Regulatory Issues Underlying Pesticide Use Patterns and Agricultural Innovation, 1987b). A recent court ruling in the United States has overturned this interpretation of the Delaney Clause, and unless the decision is reversed by a higher court a number of chemicals presently on the market will have to be withdrawn (Embers, 1993).

Pesticides approved for use on food crops must undergo testing for oncogenicity in laboratory animals. This forms the basis of the *Q star* potency factor (Q* rating) (Committee on the Scientific and Regulatory Issues Underlying Pesticide Use Patterns and Agricultural Innovation, 1987b). The Q* estimates the upper bounds of the extra tumours in humans that would be expected following exposure to a chemical in food or by other routes. The limitation in this estimation is the inherent weakness of data — obtained over a narrow range of high concentrations — when extrapolated to a wide range of very low concentrations. It is therefore used at the upper 95% confidence limits to make very conservative estimates.

While the author will go no further in reviewing the problems of cancer risk assessment, in the absence of better methods it has been in the public interest to err far on the side of safety. The results of the many years of applying these techniques suggest that the process in place has been effective. The contributions of various food types to cancer risk are shown in Table 6.

10.15 SIGNIFICANCE OF RESIDUES IN FOOD

One factor frequently overlooked by the public when confronted with the issue of residues in food is the great strides which have been made in residue detection technology. Thanks to improvements in technology, we have gone from residues reported in parts per million (ppm), to residues in parts per billion (ppb) (Kulkarni and Ashoke, 1992), and we now see reports in 0.000001 ppm, parts per trillion (ppt) (Startin et al., 1990). Examining these units in terms of time can make it easier to grasp their relative sizes: one ppm is equal to one second in 11.57 days, one ppb is equivalent to one second in 31.7 years and one ppt is equivalent to one second in 317.1 centuries. Few regulatory personnel, public officials, scientists or, for that matter, members of the general public can comprehend the significance of these numbers. Clearly, the simple detection of residues does not mean the presence of *significant* residues, let alone toxic residues. Also, let us not forget that the efficiency of government surveillance programs has improved greatly. Extreme caution is needed, therefore, when

TABLE 10.6. Estimated Oncogenic Risk Caused by the Agricultural Use of Pesticides in Food Production.

Food	Usage	Estimated Risk Number	Estimated Risk Percentage
	Herbicide		
Lettuce		1.59×10^{-4}	2.5
Chicken		1.11×10^{-4}	1.9
Beef		1.10×10^{-4}	1.9
Cottonseed		9.95×10^{-5}	1.7
Milk		5.19×10^{-5}	0.9
Tomatoes		5.17×10^{-5}	0.9
Pork		5.02×10^{-5}	0.9
Peaches		3.50×10^{-5}	0.6
Spinach		2.80×10^{-5}	0.5
Cabbage		1.82×10^{-5}	0.3
	Fungicides		
Tomatoes		8.23×10^{-4}	14.1
Oranges		3.72×10^{-4}	6.3
Apples		3.18×10^{-4}	5.4
Peaches		2.86×10^{-4}	4.9
Lettuce		1.81×10^{-4}	3.1
Potatoes		1.29×10^{-4}	2.2
Beans		1.17×10^{-4}	2.0
Grapes		1.08×10^{-4}	1.8
Wheat		6.65×10^{-5}	1.1
Celery		6.04×10^{-5}	1.1
	Herbicides		
Beef		5.38×10^{-4}	10.0
Potatoes		3.89×10^{-4}	6.7
Pork		2.17×10^{-4}	3.7
Soybeans		1.22×10^{-4}	2.1
Wheat		1.22×10^{-4}	2.1
Carrots		5.95×10^{-5}	1.0
Corn		4.98×10^{-5}	0.9
Asparagus		1.48×10^{-5}	0.3
Celery		1.04×10^{-5}	0.2
Milk		7.87×10^{-6}	0.1

Percent of total pesticide risk: insecticides 12.1; herbicides 27.1; fungicides 42.0.
Source: Committee on Scientific and Regulatory Issues Underlying Pesticide Use Patterns and Agriculture Innovation, 1987.

drawing conclusions about the possible significance of pesticide residues to man and the environment.

Finally, evidence showing the high frequency of residues in the food supply of developing countries, where control programs are not as efficient, has not to date been translated into major health problems in the exposed populations. Perhaps this suggests that the actions taken by FAO, WHO, the Codex Alimentarius Commission and its committees have served the public well. Solid evidence of the harm done by pesticide residues has been forthcoming from environmental studies; the effects on birds have been particularly obvious. Birds of prey were the sentinels which alerted the public to the developing problem, and gave us the opportunity to take measures before a more serious crisis occurred.

10.16 REFERENCES

Agriculture Canada, Agri-Food Safety Division. 1990. Annual Report 1989–1990. Ottawa, Ont.: Published by Agriculture Canada, Appendix C, 3–19.

Al-Omar, M. M. Al-Bassomy, N. Al-Ogailly, N. Shebl and D. Al-Din. 1985. "Residue Levels of Organochlorine Insecticides in Lamb and Beef from Baghdad," *Bull. Environ. Contam. Toxicol.*, 34:509–512.

Anon. 1987. Report of the National Residue Survey 1986 Results, Canberra Department of Primary Industries and Energy, Australian Government Publishing Service.

Arthur, F. C. and R. C. Axtell. 1983. "Northern Fowl Mite Population Development on Laying Hens Caged at Three Colony Sizes," *Poultry Science*, 62:424–432.

Baggot, J. 1977. "The Absorption of Drugs; Bioavailability," in *Principles of Drug Disposition in Domestic Animals*, Toronto: W. B. Saunders Co., pp. 20–41.

Baron, R. and J. Weintraub. 1986. "Immunization of Cattle against Hypodermatosis *(Hypoderma lineatum)* (Devill.) and *H. bovis* (L.) Using Lineatum Antigen," *Vet. Parasit.*, 21:43–50.

Barth, D. and J. Preston. 1985. "Efficacy of Ivermectin against the Sucking Louse Solenopotes Capillatus," *Vet. Rec.*, 116:267.

Belle, P. and A. Laval. 1981. "Deaths in Heavily Parasitized Cattle" (Translated from French), *Review de Medicine Veterainarie*, 132:63–66.

Benet, L. and L. Scheinir. 1985. "General Principles," in *The Pharmacological Basis of Therapeutics, 7th Edition*, A. F. Gilman, L. S. Goodman, T. W. Rall and F. Murad, eds., Toronto: MacMillan Publishing Co.

Borlaug, N. 1972. "Mankind and Civilization at Another Crossroad: In Balance with Nature," *Bioscience*, 22:41–44.

Bowen, H., J. Cutler and I. Graigie. 1982. "A Survey of Sheep Dip Usage in Scotland 1978," *Pesticide Science* 13(5):564–574.

Bramley, P. and D. Henderson. 1984. "Control of Sheep Scab and Other Sheep Ectoparasites with Propetamphos," *Vet. Rec.* 115:4609–463.

Busvine, J. 1976. "Pest Resistance to Pesticides," in *Pesticides and Human Welfare*, D. Gunn and J. Stevens, eds., pp. 193–205.

Campbell, J., I. Berry, D. Boxler, R. Davis, D. C. Clanton and G. Deutscher. 1987. "Effects of Stable Flies *(Diptera: Muscidae)* on Weight Gain and Feed Efficiency of Feedlot Cattle," *J. Econ. Entomol.*, 80:117–119.

Carson, R. 1961. *Silent Spring*, Greenwich, CN: Fawcett Publishing Co.

Chase, L., R. Eckerlin, J. Ebel, G. Maylin, W. Gutenmann and D. Lisk. 1987. "Residues of p,p'-DDE and Difocol in Milk of Dairy Cows Fed Commercially Produced Apple Pomace," *J. Food Safety,* 8:245–253.

Chauhan, R., Z. Singh and B. Dahiya. 1982. *First Int. Conf. Food Sci. Technol.,* Bangalore, 23–26 May, Abstract No. 9, Section II, pp. II.5.

Christensen, C., P. Nansen, S. Henriksen, J. Monrad and F. Satrijo. 1992. "Attempts to Immunize Cattle against *Ostertagia ostertagi* Infection Employing Normal and Chilled Third Stage Larva," *Vet Parasitol.,* 44:247–261.

Christie, A. 1987. "Report in Veterinary Record," *Vet. Rec.,* 121:242–243.

Codex Alimentarius Commission. 1973. "First Supplement to the List of Food Additives Evaluated for Their Safety-in-Use in Food," CAC/FAL 1–1973, Rome: FAO/WHO.

Codex Alimentarius Commission 1978. "Guide to Codex Maximum Limits for Pesticide Residues, First Issue," CAC/PR1–1978, Rome: FAO/WHO.

Codex Alimentarius Commission. 1985. "Report of the Sixteenth Session," Geneva, 1–12 July, 1985, ALINORN, Rome: FAO/WHO, 85/47.

Committee on Scientific and Regulatory Issues Underlying Pesticide Use Patterns and Agriculture Innovation. 1987a. *Regulating Pesticides in Food: Delaney Paradox,* Washington, D.C.: National Academy Press, pp. 3–44.

Committee on Scientific and Regulatory Issues Underlying Pesticide Use Patterns and Agriculture Innovation. 1987b. *Regulating Pesticides in Food: Delaney Paradox,* Washington, D.C.: National Academy Press, pp. 45–99.

Cornell University. 1972. "Pesticides, Pollution and Food Supply," Environmental Biology Report, Cornell University, 72–11.

Corrigan, P. and P. Seneviratna. 1989. "Pesticide Residues in Australian Mean," *Vet. Rec.,* 125:180–181.

Curley, A., F. Copeland and R. Kimbrough. 1969. "Chlorinated Hydrocarbon Insecticides in Organs of Stillborn and Blood of Newborn Babies," *Arch. Environ. Health,* 19:628–632.

Curtis, R. J. 1985. "Amitraz in the Control of Non-Ixodide Ectoparasites," *Vet. Parisitol.,* 18:251–264.

Davis, J. E., and R. Doon. 1987. *Silent Spring Revisited,* Washington, D.C.: American Chemical Society.

Dhaliwal, G. 1990. "Pesticide Contamination of Milk and Milk Products," in *Food Contamination from Environmental Sources,* J. Nriagu and M. Simmons, eds., New York: John Wiley and Sons, Inc., pp. 357–385.

Dickson, J., R. Peet, R. Duffy, D. Hide and D. Williams. 1983. "Heptachlor Levels in Bone Marrow of Poisoned Cattle and Horses," *Australian Vet. J.* 60:311.

Dogheim, S., M. Almaz, S. Kostandi and M. Hegazy. 1988. "Pesticide Residues in Milk and Fish Samples Collected from Upper Egypt," *J. Assoc. Off. Anal. Chem.,* 71:872–874.

Easton, E. 1987. "Evaluation of the Fly Buster, a Combination Mineral-Salt Feeder/Aerosol Spray Device for Suppression of Horn Fly *Haematobia irritans* and *Musca autumnalis," J. Agric. Entomol.,* 4:179–182.

Ecobichon, D. 1991. "Toxic Effects of Pesticides," in *Casarett and Doul Toxicology: The Basic Science of Poisons, 4th Edition,* M. Amdur, J. Doull and C. D. Klaassens, eds., New York: MacMillan Publishing Co., pp. 565–622.

Embers, L. 1993. "Move on Pesticides May Spur Food Law Change," *Chem. Eng. News,* 8:7-8.

Epstein, S. 1976. "Carcinogenicity of Heptachlor and Chlordane," *Sci. Total Environ.,* 6:103-154.

Falck, F., A. Ricci, M. Wolff, J. Godbold and P. Deckers. 1992. "Pesticides and Polychlorinated Biphenyl Residues in Human Breast Lipids and Their Relation to Breast Cancer," *Arch. of Environ. Health,* 47:143-146.

FAO. 1959. Rome: Food and Agriculture Organization, No. 1959/3.

FAO. 1966. Rome: Food and Agriculture Organization, No. PL/1965/12.

FAO. 1983. *Compilation of Legal Limits for Hazardous Substances in Fish and Fishery Products,* Rome, Italy: Food and Agriculture Organization.

FAO/WHO Panel of Experts on Pesticide Residues. 1976. *Evaluations of Some Pesticide Residues in Food,* Rome, Italy: Food and Agriculture Organization and World Health Organization.

FDA. 1990. *FDA Veterinarian,* Washington, D.C.: United States Food and Drug Administration, 5:7.

Fox, D., R. Eckerlin, J. Ebel, G. Maylin, W. Gutenmann and D. Lisk. 1989. "Chlorinated Hydrocarbon Tissue Residues from Bovine Feed Supplements: Apple Pomace and Fish Meal," *J. of Food Safety,* 9:165-172.

Frank, R., E. Smith, H. Braun, M. Holdrinet and J. McWade. 1975. "Organochlorine Insecticides and Industrial Pollutants in the Milk Supply of the Southern Region of Ontario, Canada," *J. Milk and Food Technol.,* 38:65-72.

Frank, R., H. Braun and Fleming, G. 1983. "Organochlorine and Organophosphorus Residues in Fat of Bovine and Porcine Carcasses Marketed in Ontario, Canada from 1969-1981," *J. Food Prot.,* 42:893-900.

Frank, R., H. Braun, G. Sirons, J. Rasper and G. Ward. 1985a. "Organochlorine and Organophosphorus Insecticides and Industrial Pollutants in the Milk Supplies of Ontario—1987," *J. Food Prot.,* 48:499-504.

Frank, R., J. Rasper. H. Braun and G. Ashton. 1985b. "Disappearance of Organochlorines Residues from Abdominal and Egg Fat of Chickens, Ontario Canada 1969-1982," *J. Assoc. Off. Anal. Chem.,* 65:124-129.

Frank, R. and B. Ripley. 1990. "Food Residues from Pesticides and Environmental Pollutants in Ontario," in *Food Contamination from Environmental Sources,* J. Nriagu and M. Simmons, eds., New York: John Wiley and Sons, Inc., pp. 473-524.

Freis, G. 1972. "PCB Residues: Their Significance to Animal Agriculture," *Agr. Sci. Rev.,* USDA, Third Quarter, pp. 19-24.

Fries, G., G. Morrow and C. Gordon. 1973. "Long-Term Studies of Residues and Excretion by Cows Fed a Polychlorinated Biphenyl (Aroclor 1254)," *J. Agr. Food Chem.,* 21:117-120.

Goldstein, A., L. Aronow and S. Kalman. 1974. "Drug Absorption, Routes of Administration," in *Principles of Drug Action, The Basis of Pharmacology,* New York, Toronto: J. Wiley and Sons, Inc., pp. 143-149.

Grimson, R., R. Stilborn, P. Gummeson, W. Leaning, J. Guerrero and K. Newcomb. 1987. "Effect of Antiparasitic Treatments on the Performance and Profitability in Feedlot Steers," *Modern Vet. Practice,* 68:361-364.

Gunn, D. and J. Stevens. 1976. *Pesticides and Human Welfare,* Oxford University, Great Britain: Oxford University Press, pp. 88-91.

Hall, R., J. Vanderpoperliere, F. Fisher, J. Lyons and K. Doisy. 1983. "Comparative Efficacy of Plastic Strips Impregnated with Pyrethrin and Permetherin Dust for Northern Fowl Mite Control on Caged Laying Hens," *Poultry Sci.*, 62:612–615.

Harradine I. and K. McDougall. 1986. "Residues in Cattle Grazed on Land Contaminated with Heptachlor," *Australian Vet. J.*, 63:419–422.

Hayes, W., Jr. 1975. *Toxicology of Pesticides*, Baltimore: Williams and Wilkins, pp. 322–326.

Hayes, W., Jr. 1982. *Pesticide Studies in Man*, Baltimore: Williams and Williams, pp. 593–594.

Health and Welfare Canada. 1988. "Regulations for Pesticide Residues in Food," in *Food and Drug Act*, Updated to Dec/88.

Hecht, H. 1985. "Ruckstande in Fleisch und ihre Problematik," *Fleischwirtschaft*, 67:797–805.

Hewitt, G. and T. Heard. 1982. "Phosmet for the Systemic Control of Pig Mange," *Vet. Rec.*, 111:558.

Hogsette, J. and J. Ruff. 1986. "Evaluation of Flucythrinate and Fenvalerate Impregnated Ear Tags and Permethrin Ear Tags for (Diptera: Muscida) Control on Beef and Dairy Cattle in North West Florida," *J. Econ. Entomol.*, 79:152–157.

Holmstedt, B. 1963. "Cholinesterases and Anticholinesterase Agents," in *Handbuch der Experimentellen Pharmakologie, Vol. 15*, G. Koelle, ed., Berlin: Springer-Verlag, pp. 428–485.

Hopkins, T., I. Woodley and R. Blackwell. 1985. "The Safety and Efficacy of Flumethrin Pour-on Used to Control *Boophilus microphilus* on Cattle in Australia," *Vet. Med. Rev.*, 2:112–125.

Ibrahim, K. and M. Abu-Samra. 1987. "Experimental Transmission of a Goat Strain of Sarcopotes Scabiei to Desert Sheep and Its Treatment with Ivermectin," *Vet. Parasit.*, 26:157–164.

Kalra, R. and R. Chawla. 1986. "Pesticide Contamination of Foods in the Year 2000 A.D.," *Proc. Indian Natl. Sci. Acad.*, B52:188–204.

Kaphalia, B., F. Siddiqui and T. Seth. 1985. "Contamination Levels in Different Food Items and Dietary Intake of Residues in India," *Indian J. Med. Res.*, 81:71–78.

Kapoor, S., R. Chawla and R. Kalra. 1980. "Contamination of Bovine Milk with DDT and HCH Residues in Relation to Their Usage in Malaria Control Programme," *J. Environ. Sci. Health*, B15:545–557.

Kerbabaev, E., V. Kuznetsov, L. Bodreeva and L. Lekanova. 1987. "Aerosol of the Insecticide Etafos for Controlling Flies in Pig Houses," *Veterinariya Moscow*, 9:23–24.

Khandekar, S., A. Noronha and S. Banerji. 1981. "Organochlorine Pesticide Residues in Eggs and Milk Available in Bombay Markets," *Sci. Cult.*, 47:137–139.

Khan, M. and G. Korzub. 1981. "Systematic Control of Cattle Grubs *(Hypoderma* spp.*)* in Steers Treated with Warbex and Weight Gains Associated with Grub Control," *Can. J. Comp. Med.*, 45:15–19.

Kulkarni, A. and M. Ashoke. 1990. "Pesticide Contamination of Foods in the United States," in *Food Contamination from Environmental Sources*, J. Nriagu and M. Simmons, eds., New York: John Wiley and Sons, Inc., pp. 257–293.

Lauger, P., H. Martin and P. Muller. 1944. "Uber Konstitution und Toxischee Wirkung von Naturlichen und Neuen Synthetischen Insektentotenden Stoffen," *Helv. Chim. Acta.*, 27:892.

Leibisch, A. 1986. "Bayticol Pour-on: A New Product and a New Method for the Control of Stationary Ectoparasites in Cattle," *Vet. Med. Review,* 1:17–27.

Maderena, G., G. Dazzie, C. Campanini and E. Maggie. 1980. "Organochlorine Pesticide Residues in Meat of Various Species," *Meat Science,* 4:157–166.

Mazariegos, F. 1976. "Impact Monitoring of Agriculture Pesticides," United Nations Environmental Programme, AGP:1796/M4, Rome, pp. 15–18.

Mehrotra, K. 1985a. "Use of DDT and Its Environmental Effects in India," *Proc. Indian Natl. Sci. Acad.,* B51:169–184.

Mehrotra, K. 1985b. "Use of HCH (BHC) and Its Environmental Effects in India," *Proc. Indian Natl. Sci. Acad.,* B51:551–595.

Mehrotra, K. 1986. "Pest Control Strategies for 2000 A.D.," *Proc. Indian Natl. Sci. Acad.,* B52:10–16.

Meleny, W., F. Wright and F. Guillot. 1982. "Residual Protection against Cattle Scabies Afforded by Ivermectin," *Am. J. Vet. Res.,* 43:1767–1769.

Menzer, R. E. 1991. "Water and Soil Pollutants," in *The Basic Science of Poisons, Casarett and Doull's Toxicology,* New York: MacMillan Publishing Co., Chapter 26, pp. 872–902.

Mercer, D., R. Teske and J. Condon. 1976. "Herd Health Status of Animals Exposed to Polychlorinated Biphenyl (PBB)," *J. Toxicol. Environ. Health,* 2:335–349.

Muan, B., J. Skare, N. Soli, K. Grave and S. Odegaard. 1985. "Use of Malathion against Ectoparasites on Lactating Cows," *Acta. Vet. Scandinavica,* 26:352–362.

NCI. 1979. NCI Carcinogenesis Technical Report Series No. 37, Publication No. (NIH) 79-837 NCI, Bethesda, MD.

National Animal Poison Control Center. 1986. Annual Report. Urbana, IL.

National Research Council. 1984. *Cancer Today: Origins, Preventions and Treatment,* Washington, D.C.: National Academy Press.

Narahashi, T. 1985. "Nerve Membrane Ionic Channel as the Primary Target of Pyrethroids," *Neurotoxicology,* 6:3–22.

Narahashi, T. 1986. "Mechanism of Action of Pyrethroids on Sodium and Calcium Channel Gating," in *Neuropharmacology and Pesticide Action,* M. Ford, G. Lunt, R. Reay and P. Usherwood, eds., Chichester, England: Ellis Horwood, pp. 36–60.

Neumann, G. 1988. "The Occurrence and Variation of Organochlorine Pesticides Residues Detected in Australian Livestock at Slaughter," *Acta. Vet. Scandinavica,* Suppl. 84:299–302.

Nijhuis, H. and C. Ewers. 1984. "Detection and Determination of Coumaphos and Coxoron in Milk," *Milchwissenschaft,* 41:133–135.

O'Brien, R. 1967. *Insecticides, Action and Metabolism,* New York: Academic Press.

Ontario Ministry Agriculture and Food. 1989–90. *Field Crop Recommendations: Pesticides,* Publication No. 296, Toronto, Ontario: Ontario Ministry of Agriculture and Food, 75–82.

Pennington, J. and E. Gunderson. 1987. "History of the Food and Drug Administration Total Diet Study," *J. Assoc. Anal. Chem.,* 70:772–782.

Petterson, D., R. Casey, G. Bell and J. McIntyre. 1988. "Residues in Beef Cattle Accidently Exposed to Commercial Heptochlor," *Aust. Vet. J.,* 64:50–53.

Quisenberry, S. and D. Foster. 1984. "Cost Benefit Evaluation of House Fly *(Diptera: Muscidae)* Control in Caged Layer Poultry Houses," *Poultry Sci.,* 63:2132–2139.

Salathe, L. and R. Buse. 1979. "Household Food Comsumption Patterns in the United States," *U.S. Department of Agriculture Technical Bulletin*, 1587.

Saschenbrecker, P. 1976. "Levels of Terminal Pesticide Residues in Canadian Meat," *Can. Vet. J.*, 17:158–163.

Saxena, M. and M. Siddiqui. 1982. "Pesticidal Pollution in India: Organochlorine Pesticides in Milk of Women, Buffalo and Goats," *J. Dairy Sci.*, 65:430–434.

Singh, P., R. Battu, B. Joia, R. Chawla and R. Karla. 1986. "Contribution of DDT and HCH Used in Malaria Control Programme Towards the Contamination of Bovine Milk," in *Proc. Symp. Pesticide Residues and Environmental Pollution*, Muzaffarnagar, India: S. D. College, pp. 86–92.

Skoda, S., J. Campbell and S. Kunz. 1987. "Wide Area Treatment of Cattle for Hornflies and Face Flies *(Diptera:Muscidae)* in South-Central Nebraska," *J. Econ. Entomol.*, 80:811–816.

Smith, R. and D. Lewis. 1988. "A Potpourri of Poisonings in Alberta in 1987," *Vet. Human Toxicol.*, 30:118–120.

Sossa, E. 1985. "Evaluatikon of the Efficacy and Residual Effect of Flumethrin Pour-On against *Boophilus microphilus* in Cattle in Uraguay," *Vet. Med. Rev.*, 2:126–131.

Startin, J., M. Rose, C. Wright, I. Parker and J. Gilbert. 1990. "Surveillance of British Foods for PCDD's and PCDF's," *Chemosphere*, 20:793–798.

Steffy, K., J. Mack, C. Macmonegle and H. Petty. 1984. "A 10 Year Study of Chlorinated Hydrocarbon Insecticide Residues in Bovine Milk in Illinois (USA) 1972–1981," *J. Environ. Sci. Health*, B19:49–66.

Tarry, D. W. 1981. "Distribution of Cattle Warble Flies in Britain: A Larval Survey," *Vet. Rec.*, 108:69–72.

Taylor, P. 1990. "Anticholinesterase Agents," in *The Pharmacological Basis of Therapeutics, 7th Edition*, A. Gilman, L. Goodman, T. Rall and F. Murad, eds., New York: MacMillan Publishing Co., pp. 131–149.

Tonogai, Y., Y. Hasegawa, Y. Nakamura and S. Fujino. 1989. "Simultaneous Determination of Ten Kinds of Organochlorines Insecticides in Beef by Gas Chromatography," *J. Food Protection*, 52:92–95.

Travis, C., A. Arms and C. Gartenl. 1987. "Bioconcentration of Organics in Beef, Milk and Vegetation," Report to U.S. Government under Contract DE-AC05-840R21400, 1–15.

UNEP, FAO, WHO. 1989. "Guidelines for Predicting Dietary Intake of Pesticides," UNEP, FAO, WHO, Food Contamination Monitoring Programme, World Health Organization, Geneva, ISBA, 92,4,154250,0.

USFDA FSIS. 1988. *Domestic Residue Data Book: National Residue Program 1987*, Washington, D.C.: United States Department of Agriculture, Food Safety and Inspection Service.

USFDA FSIS Science Program. 1989. "Compound Evaluation and Analytical Capability," National Residue Program Plan, Washington, D.C.: United States Department of Agriculture.

Uyeta, M., S. Taue and T. Nishimoto. 1970. "Residues of BHC Isomers and Other Organochlorine Pesticides in Fatty Foods of Japan," *J. Fd. Hyg. Soc.*, 11:256–263.

Venant, A. and L. Richor-Bac. 1985. "Example of the Contamination of Sheep Fat by Lindane," *Bulletin Mensuel de la Societe Veterinasire Pratique de France*, 69:363–378.

Webster, K., M. Rankin, N. Goddard, D. Tarry and J. Coles. 1992. "Immunological and Feeding Studies on the Antigens Derived from the Biting Fly *Stomoxys calcitrans*," *Vet. Parasitol.*, 44:143–150.

Welborn, J., R. Allen, G. Byker, A. DeGow, J. Hertel, R. Noordhock and D. Koons. 1975. Report from Senate Special Investigating Committee, Lansing, MI.

Whiting, F., O. Lough and W. Brown. 1987. "Occurrence of DDE in Dairy Feeds in the Arizona Milk Shed," *Bulletin Environ. Contam.*, 39:587–592.

WHO International Task Group on Environmental Health Criteria for DDT. 1989. *DDT and Its Derivatives—Environmental Aspects*, Geneva: Published Jointly by United Nations Environment Programme, the International Labour Organization and World Health Organization, pp. 9–78.

Wright, F. C. 1986. "Control of Psoroptic Scabies of Cattle with Fenvalerate," *Vet. Parasit.*, 21:37–42.

CHAPTER 11

Herbicides

JOSEPH V. KITZMAN, D.V.M., Ph.D.[1]

11.1 INTRODUCTION

D RAMATIC increases in crop yields throughout the world in modern times have been due, to a great extent, to effective pest control. Controlling pests, including weeds, insects, fungi, bacteria, and rodents, has largely been accomplished by direct application of chemical agents. The widespread use of pesticides (insecticides, herbicides, and fungicides) has brought about both benefit and risk to humans, animals and the environment.

Although herbicide application may cause direct acute and chronic toxicity, herbicide risk may be increased during the dissemination process. Application methods may result in herbicide contact with nearby nontarget plants and animals which ultimately adds to animal and human exposure levels (Dover, 1985). Additionally, contamination of drinking water has become an increasingly significant source of exposure.

The purpose of this chapter will be to examine herbicide use and the resulting risk to consumers. The most commonly used herbicides will be examined to assess toxicity and cancer-causing potential. The processes of registration and regulation of herbicide use will be examined. Finally, case reports of accidental or experimental animal contact with herbicides will be reviewed so that the risk-benefit aspect of herbicide use can be more clearly evaluated.

[1]Chairman, Department of Basic and Applied Sciences, College of Veterinary Medicine, Mississippi State University, Mississippi State, Mississippi, USA.

11.2 HERBICIDES—WHAT ARE THEY?

Weed control is an essential part of efficient farming. Herbicides, espe-cially those approved in the past 10 years, are widely used in the U.S. to enhance the yield of a variety of cultivated crops, pastureland, and harvested fruits and vegetables. Herbicide use has increased markedly in the past 15 years because the efficient use of herbicides has proven to be a positive economic factor for farmers. Over 90% of U.S. acreage of corn, soybeans, rice, and peanuts were treated with herbicides in 1982. 71% of tobacco acreage, 59% of grain sorghum, and approximately 40% of wheat, barley, and oats were treated with herbicides (Tshirley, 1987). Herbicides used for weed control on corn and soybean acreage are listed in Table 11.1. Put into perspective, U.S. herbicide use has increased from 98 million kg in 1971, to 171 million kg in 1976, to 198 million kg in 1984 (Delvo, 1984). Herbicides account for approximately 65% of pesticide use on farms, with fungicide and insecticide use making up the remaining 35% of farm chemical usage.

Currently, over 120 chemicals are approved for use as herbicides on one or more crops in the United States. Inherent in this group of chemicals are differences in phytotoxicity, selectivity, mode of action, persistence, use, and risk. When a new chemical is evaluated as a potential herbicide, it is initially identified with an existing class based on chemical structure (Mrak, 1974). Because of its structure, the chemical would be expected to share potential environmental or health-related effects with other chemi-cals in this class of herbicides. Also, if a currently approved herbicide was suspected of having a particular health-related effect, a chemical classifica-tion system would be useful for selecting other chemically related herbi-cides for subsequent examination and testing. Table 11.2 presents a classification of herbicides based on chemical similarities, and includes one or more commonly used herbicides in each class. Brief descriptions of each class are presented, including animal toxicities and oncogenicity, if known.

11.2.1 INORGANIC HERBICIDES

The inorganic herbicides are generally water soluble and are applied as water solutions, oil-water emulsions, or in dry form. They are not retained in the soil, and are used at very low application rates. These chemicals, in-cluding sodium chlorate and copper sulfate, are toxic to fish, and may be toxic to mammals if relatively large quantities are ingested acutely (5,000 mg/kg—sodium chlorate) (Beste, 1983). Cases of chronic toxicity are un-

TABLE 11.1. Herbicides Used on Corn and Soybeans during 1982.

Herbicide	Active Ingredient (million kg)	
	Corn	Soybeans
Single applications		
Acifluorfen		0.9
Alachlor	8.9	10.3
Atrazine	1.0	
Bentazon		6.7
Butylate	1.0	
Chloramben		2.7
Cyanazine	2.2	
2,4–D	1.4	
Dicamba	0.4	
Fluchloralin		2.6
Glyphosate		2.2
Linuron		1.3
Metolachlor	1.4	6.9
Metribuzin		2.2
Trifluralin		20.4
Other	4.2	5.5
Total	39.0	61.7
Tank mixes		
Acifluorfen + bentazon		0.3 + 0.7
Alachlor + metribuzin		6.9 + 1.7
Alachlor + linuron		8.1 + 3.2
Alachlor + naptalam + dinoseb		1.5 + 1.3 + 0.6
Atrazine + alachlor	7.4 + 9.5	
Atrazine + butylate	3.9 + 1.2	
Atrazine + cyanazine	1.2 + 1.6	
Atrazine + metolachlor	3.9 + 4.8	
Atrazine + simazine	.58 + .54	
Bentazone + 2,4–D		0.4 + *
Chloramben + alachlor		1.5 + 1.8
Chloramben + trifluralin		0.9 + 0.5
Cyanazine + alachlor	2.7 + 3.4	
Cyanazine + butylate	1.2 + 2.2	
Cyanazine + metolachlor	0.4 + .54	
Dicamba + 2,4–D	.45 + .72	
Dinoseb + naptalam		1.2 + 2.4
Metolachlor + metribuzin		4.2 + 1.6
Metolaclor + atrazine + simazine	1.4 + 1.1 + .58	
Metolachlor + cyanazine + atrazine	.63 + .27 + .27	
Oryzalin + linuron		0.4 + 0.3
Paraquat + others		0.4 + 1.7
Trifluralin + metribuzin		8.8 + 3.8
Other	4.2	11.1
Total	64.6	65.3
Total herbicides	103.6	127.0

*Less than 100,000 pounds.
Source: Tshirley (1987).

TABLE 11.2. Herbicide Chemical Classification.

Chemical Class	Examples
Inorganic Herbicides	Sodium chlorate, copper sulfate
Organic Herbicides	
I. Aliphatics	
A. Chlorinated acids	Dalapon
B. Organic arsenicals	Methanearsonic Acid
C. Others	Glyphosate
II. Amides	
A. Chloroacetamides	Alachlor, Metolachlor
B. Others	Pronamide
III. Benzoics	Dicamba
IV. Bipyridilliums	Paraquat
V. Carbamates	Asulam
VI. Dinitroanilines	Oryzalin, Trifluralin
VII. Diphenyl Ethers	Acifluorfen, Diclofop
VIII. Nitriles	Bromoxynil
IX. Phenoxys	2–4D
X. Thiocarbamates	Diallate, Molinate
XI. Triazines	Atrazine, Terbutryn
XII. Uracils	Bromacil
XIII. Ureas	Linuron
XIV. Unclassified	Oxadiazon, Picloram

known. The inorganic herbicides have not been classed as oncogenic by EPA.

11.2.2 ORGANIC HERBICIDES

11.2.2.1 Aliphatics

The aliphatic herbicides can be divided into the subclasses of chlorinated acids, organic arsenicals, and an unclassified category, which includes the widely used herbicide glyphosate. Chlorinated acids, including dalapon, are soluble in water and are generally applied as pre- or early postemergent agents to control annual or perennial grasses. They exhibit a low order of acute and chronic toxicity to fish and mammals. Dalapon has caused irritation to skin upon repeated exposure; however, it is not absorbed through the skin in significant amounts. The group of compounds is currently classed as nononcogenic.

Organic arsenical herbicides are derivatives of methanearsonic acid. All formulations are 100% water soluble, and are used as postemergent herbicides for broadleaf weed control in cotton fields. These compounds show little, if any, acute or chronic toxicity to fish or mammals. The herbicide methanearsonic acid has been classed as potentially oncogenic by EPA.

Glyphosate is a very broad spectrum, water soluble herbicide widely used as a postemergent and for weed control around fields and ponds. The chemical is moderately toxic to fish (trout and bluegill) but is essentially nontoxic to birds and mammals. Glyphosate has been extensively used in the U.S. since its introduction in 1971, and has been classed as potentially oncogenic by EPA.

11.2.2.2 Amides

Herbicides in the amides class contain several chemicals that exhibit significant solubility in solvents, and are sparingly soluble in water. They are used as preemergent and postemergent herbicides on corn and soybeans, among other crops. This class is moderately toxic to fish, but relatively nontoxic to birds or mammals. Special handling instructions, including protective clothing and the use of gloves, should be followed by operators. Alachlor was not found in runoff water from the intensively farmed Black Creek Watershed, Indiana, in the pesticide and PCB residue study performed by EPA (Dudley and Kerr, 1980). Herbicides in this class, including alachlor, metolachlor, and pronamide have been determined to be potentially oncogenic by EPA.

11.2.2.3 Benzoics

The benzoic herbicides are comprised of low molecular-weight aromatic chemicals formed by attaching chlorine and other simple groups to the benzoic acid molecule. The benzoics are soluble in alcohols and organic solvents, and only sparingly soluble in water. This group of herbicides is relatively nontoxic, and has been used as a preemergent and postemergent application treatment alone and in combination with other chemicals (e.g., dicamba + 2,4–D). Following a 5-day oral administration of 2.2 mg/kg dicamba daily in the cow, the animal excreted 89% (urine) and 2% (feces) of the administered dose within 6 hours following administration. Essentially no dicamba was excreted in the milk (Oehler and Ivie, 1980). The EPA has not classified this group of herbicides as potentially oncogenic.

11.2.2.4 Bipyridiliums

The major herbicides of the bipyridilium group, paraquat, has been used for weed control in grass seed crops, orchards, and in rice and soybean cultivation. This water-soluble chemical is nontoxic to wildlife, fish, and mammals when applied at recommended rates. Paraquat has been shown to be toxic only when administered in the mg/kg body weight range (Clark et al., 1966) with the LD_{50} in turkeys being about 20 mg/kg (IV injection), 100 mg/kg (IP injection), 290 mg/kg (oral), and 500 mg/kg (dermal application) (Smalley, 1973). Paraquat has been classed as potentially oncogenic by EPA.

11.2.2.5 Carbamates

Asulam, the major carbamate herbicide, is used primarily to control johnsongrass in sugarcane fields. Asulam is water-insoluble, therefore the sodium salt, aqueous solution is broadcast by ground or air application. Toxicity to wildlife, fish, and mammals is negligible. The EPA has identified asulam as potentially oncogenic.

11.2.2.6 Dinitroanilines

Dinitroaniline derivatives have been used since 1963, when the herbicidal properties of trifluralin were reported. These chemicals are poorly soluble in water and are generally used as preemergents in the soil. Although toxic to fish if sprayed directly in water, these herbicides should not be a hazard if applied by recommended methods. Acute and chronic feeding trails are well tolerated by laboratory animals. Oryzalin and trifluralin have been identified as potentially oncogenic.

11.2.2.7 Diphenyl Ethers

The diphenyl ether herbicides are used primarily as postemergent (2- and 4-leaf stage) agents in soybean fields. These chemicals are water-insoluble, and represent significant risk to handler and environment. Full protective clothing, including goggles and respirator, should be worn during handling, mixing, and during application. These agents are toxic to fish via direct or runoff contamination of lakes, ponds, or streams. Corneal opacity and skin irritation have been demonstrated in the rabbit. EPA has identified acifluorfen and diclofop as potentially oncogenic.

11.2.2.8 Nitriles

The nitriles are water-insoluble, postemergent herbicides used on wheat, barley, oats, and along roadways and rights-of-way to control certain broadleaf weeds. This class of herbicides is toxic to fish, and produces skin and conjunctival irritation in rabbits. The nitriles are not currently classed as oncogenic by EPA.

11.2.2.9 Phenoxys

The phenoxyalkanoic acids, including 2,4–D, have maintained an important position in the herbicide market. The chemicals are soluble in organic solvents, and are applied as amine or ammonium salts. These herbicides have been used on crops and pastures for broadleaf weed control, and may be used for brush control at higher application rates. 2,4–D is toxic to fish, and has demonstrated acute oral toxicity in the range of 300–1,000 mg/kg in rats, guinea pigs, and rabbits (Beste, 1983). Skin irritation has been reported in the rabbit. Pharmacokinetic studies in rats and dogs (Leng, 1977), as well as sheep and cattle (Leng, 1972), have indicated that residues in tissues would only be likely if humans consumed meat from animals slaughtered while ingesting freshly treated forage or pastures treated as high application rates. Very recently, the EPA has identified 2,4–D as potentially oncogenic.

11.2.2.10 Thiocarbamates

The thiocarbamates exhibit varying water solubility, with newer products being more water soluble (molinate) than older ones (diallate). The herbicides are applied for postemergent control of weeds in a variety of crops including corn, potatoes, soybeans, and rice. This class of chemicals is relatively nontoxic to fish, wildlife, and mammals. One herbicide in the class of thiocarbamates, diallate, is currently classified as potentially oncogenic by EPA.

11.2.2.11 Triazines

Triazine herbicides are a group of water-insoluble chemicals used for control of broadleaf and grassy weeds in numerous crops such as corn, sorghum, barley, wheat, and on rangeland. The triazines vary from low to moderately toxic to fish, but are relatively nontoxic to wildlife and mammals. Atrazine was not found to persist in runoff water from the Black Creek Indiana Watershed (see section 11.2.2.2) (Dudley and Kerr, 1980). The EPA has classified atrazine and terbutryn as potentially oncogenic.

11.2.2.12 Uracils

The uracils are wettable powder derivations of water-insoluble chemicals. These herbicides, including bromacil, are used to control grasses and broadleaf weeds primarily in noncropland areas, such as orchards. The uracil herbicides are slightly toxic to fish, and exhibit very low toxicity potential to wildlife and mammals. No herbicides in this class are currently classified as oncogenic by EPA.

11.2.2.13 Ureas

The urea herbicides are used to control germinating and newly established broadleaf weeds in several crops and along roadsides. However, the most extensive use of these water-insoluble chemicals is on soybeans. During acute laboratory experiments, the urea herbicides were quite nontoxic to fish, wildlife, and mammals. Linuron, the most extensively used herbicide in this class, has produced increases in testicular tumors during chronic toxicity studies. Based on oncogenic potential and usage, the EPA has identified linuron as the approved herbicide with the highest overall oncogenic potential.

11.2.2.14 Unclassified

The unclassified herbicides do not readily fit into any of the other classes due to dissimilar chemical structure, mode of action, or use. Currently, more than 20 herbicides are grouped in the unclassified category. Only one herbicide in this group, oxadiazon, has been identified by EPA as potentially oncogenic.

11.3 EPA TOXICOLOGY REQUIREMENTS FOR HERBICIDE REGISTRATION

To protect environmental safety associated with herbicide use, the EPA has established testing guidelines for herbicide registration. Significant data requirements regarding herbicide toxicity to nonhuman organisms, and the extrapolation of these data to human safety, have been established (EPA, 1982, 1984).

Basic toxicology studies required by the EPA are shown in Table 11.3. These studies are divided into acute, subchronic, chronic, mutagenicity, and special classifications, and are conducted primarily in the rat.

Acute toxicity determination is the first step of herbicide safety evaluation. The effects of ingestion, inhalation, skin and eye contact are evaluated for a 2 to 3 week period following a single-dose exposure. If repeated dermal exposure to a chemical listed in Table 11.3 is anticipated, dermal sensitization testing must be performed. The acute toxicity test is used to establish herbicide toxicity in relation to other chemicals, and determines the median lethal dose for each route of administration.

Following acute toxicity determination, the herbicide is evaluated in

TABLE 11.3. EPA Toxicology Data Requirements.

Test	Species
ACUTE	
Oral/Dermal/Inhalation	Rat, Rabbit
Primary Eye/Dermal Irritation	Rabbit
Dermal Sensitization	Guinea Pig
Delayed Neurotoxicity	Hen
SUBCHRONIC	
90 Day Feeding	Rat and Dog
90 Day Dermal/Inhalation	Rat
90 Day Neurotoxicity	Hen
CHRONIC	
Oncogenicity	Rat and Mouse
Chronic Feeding	Rat and Dog
Teratogenicity	Rat and Rabbit
Reproduction, 2–Generation	Rat
SPECIAL	
Metabolism	Rat
Dermal Penetration	Rat

Source: Cardova, R.A. 1987. Current Toxicology Requirements for Registration. IM, Pesticides—Minimizing the Risk, ACS Symposium Series, 1987.

subchronic tests. Subchronic toxicity testing usually involves daily exposure to the chemical for up to 90 days in species (rat and dog), at several dosage levels. The animals in these studies are monitored daily for overt signs of toxicity, change in appetite, weight loss, death, and histopathologic changes in body organs. Chronic toxicity testing is intended to assess long-term exposure to herbicide chemicals, usually at three dosage levels. The capacity of the chemical to produce benign or malignant tumors is evaluated by lifetime oncogenicity studies in rats and mice. Chronic feeding studies must also be performed in rodent and nonrodent species (usually the dog) to satisfy EPA's oncogenicity assessment. Additionally, one- and two-generation feeding trials are usually required in the rat and rabbit to assess the effects of the herbicide on reproduction, including gonadal function, mating behavior, conception, and teratogenicity.

Mutagenicity studies are performed to determine if a herbicide is capable of damaging DNA or producing other gene mutations. These mutagenicity studies are important in the overall evaluation of the chemical because of their relation to carcinogenesis. It is generally accepted that the initial step of tumor formation is a genetic mutation event, thereby linking mutagenicity to carcinogenicity.

Other tests include a metabolism study in rodents to determine fate of the herbicide in the animals. Rates of absorption, distribution, biotransformation, and excretion can be calculated at one or more dosage rates. From these parameters, estimates of biological half-life, elimination, and persistence in the body can be determined. If dermal penetration is a likely route for human exposure, dermal absorption tests in rodents may be required for approval.

The purpose of toxicity testing is to provide the EPA with data to evaluate the hazard and assess the risk associated with using a particular herbicide. Generally, the acceptable daily intake of a chemical for humans incorporates a safety factor of 100 times less than the no-observable-effect level (NOEL) in the most sensitive animal test species. Carcinogenic or mutagenic chemicals are approved for use based only on their perceived value and risk to society.

11.4 MONITORING HERBICIDE RESIDUE IN FOODS

The U.S. government, through several cooperating agencies, bears the responsibility for protecting the public from the potential risk caused by illegal residues of chemicals, including herbicides, in our food supply. These agencies are the Environmental Protection Agency (EPA); the Food and Drug Administration (FDA) of the Department of Health and Human Services; and the U.S. Department of Agriculture (USDA), particularly involving the Food Safety and Inspection Service (FSIS).

As outlined earlier, the EPA regulates the manufacture, distribution, and use of herbicides by enforcing the Federal Insecticide, Fungicide, and Rodenticide Act (FIFRA), originally enacted in 1947. To receive EPA registration, a herbicide must accomplish its intended use without adversely affecting human health or the environment. EPA also establishes tolerances for herbicides in the food supply involved in interstate shipment under the Federal Food, Drug, and Cosmetics Act (FFDCA). Tolerance limits are established whenever the approved use of a herbicide results in a known residue in food.

As a part of the registration and regulation of herbicides, the EPS maintains a data management system for storing detailed information about each chemical. This data storage system, called the EPA Tolerance Assessment System (TAS), contains information for basic toxicologic assessment as well as projecting chronic effects due to repeated exposure. Each TAS product file contains the herbicide name and its specific classification code, the crop or crops for which it is approved for use, each application rate, and the most currently approved tolerance level.

The TAS also maintains food consumption data based on a USDA survey of over 30,000 individuals, dividing consumed food into 376 unique types (Rizek, 1977–78). Since some of these food groups contain commodities which are known to contain herbicide residues, the EPA can predict intake of each chemical in the various food groups. More importantly, the EPA can predict risk to human health, including carcinogenicity, following long-term exposure to herbicides and other farm-use chemicals.

The EPA currently registers approximately 290 active pesticide ingredients for agricultural use. Based on field testing to establish residue levels, the EPA has established over 7,400 food tolerance levels to cover all expected food residues from the approved group of pesticides. These food tolerances include residues on raw commodities and processed foods.

Herbicides represent a significant component of EPA's pesticide regulation and tolerance-setting activity. EPA has approved more than 120 herbicides for agricultural use, and has established about 2,500 food tolerances for herbicides in raw and processed foods.

The FDA bears the responsibility for ensuring that foods intended for human consumption (other than meat, poultry and associated products) are safe and do not contain illegal residues of chemicals or environmental contaminants. Annually, the FDA samples about 7,000 domestic and 5,000 imported food products to monitor residue of herbicides, fungicides, and insecticides (FDA, 1986).

The USDA is responsible for assuring wholesomeness and safety of meat, poultry, and associated products intended for human consumption through enforcement of the Federal Meat Inspection Act, the Poultry Products Inspection Act, and the Egg Products Inspection Act. Within the USDA, FSIS is responsible for sampling and analyzing meat and poultry products to determine if violative amounts of chemicals are present. Because chemical residues in meat and poultry products pose a potential threat to public health, the National Residue Program (NRP) of FSIS monitors residues by randomly sampling meat and poultry products at domestic slaughter facilities and at official ports of entry receiving imported products. Results of NRP monitoring help identify sources of adulterated meat and poultry products and stop movement and distribution of such products.

11.4.1 HERBICIDE RESIDUES IN FOODS

Approximately 120 herbicides are currently registered for use on one or more crops in the U.S., and these herbicides are applied at the rate of 230 kg/year. Since herbicides are used to control broadleaf weeds associated with the production of grain crops such as corn and soybeans, residues of the herbicides could be ingested by animals intended for slaughter when fed these grains as feedstuffs. However, the risk of toxicity to humans from the herbicides is generally low when the herbicide is applied at recommended application rates, because of rapid chemical degradation following spraying, and rapid elimination from the body if the herbicide is ingested.

Of all toxicities produced by herbicides, the oncogenic risk is the most serious, and, by far, the most publicized. The EPA uses the Delaney Clause (a provision of the Food, Drug, and Cosmetic Act) for granting approval and establishing tolerance for suspected oncogenic herbicides. The EPA sets tolerances for raw and processed agricultural products using different procedures, reflecting EPA's concern for oncogenic risk assessment.

The tolerance level for a suspected carcinogenic herbicide in or on a raw agricultural product is established by estimating quantitative risk assessment to the human population. Lifetime exposure to the herbicide is estimated by the amount of the herbicide in the food product, lifetime food product ingestion estimates, and the oncogenic potency of the herbicide (a measure of tumor incidence). On the basis of this assessment, the EPA decides whether a specific tolerance for a particular herbicide poses a significant risk to humans. If the risk is not determined to be significant, the EPA will permit use of the oncogenic chemical based on its benefits to society.

Establishing tolerances for herbicides in processed agricultural products (dried, cured, canned, etc.) is a more complex assessment. If residues of an oncogenic herbicide are present in processed food, but do not concentrate during the processing step, the EPA can assess risk, safety, and benefit, and may permit use of the herbicide. However, if the oncogenic herbicide concentrates in the processed food, the EPA automatically denies the request without further examination of risk assessment.

Currently, the EPA does not recognize any processed food forms of meat, milk, or poultry products. Consequently, herbicide concentration during processing is not an issue in these food products at this time. However, processed by-products of several crops are fed to animals intended for slaughter (Table 11.4). If any of these by-products should prove to concentrate oncogenic herbicides during the processing phase, the EPA would probably not permit their continued use as feed to animals.

The impact of the Delaney Clause would be even more pronounced if the EPA determined that animal feed processing on the farm, such as drying, curing, or silage production, resulted in concentration of oncogenic herbicides. Currently, the EPA does not consider these activities as processing. Table 11.5 lists current animal feeds not subject to processed feed regulations.

The oncogenic risk following repeated exposure to certain herbicides is a continuing problem of risk-benefit assessment and regulation. The EPA has identified 18 potentially oncogenic herbicides used in the U.S. at the

TABLE 11.4. Processed By-Products Used as Animal Feed That May Require Tolerances.

	Estimated Waste Used for Livestock Feed	
Commodity	Percentage	Metric Ton, Wet
Apricots	47	4,897
Asparagus	44	15,272
Bananas		
Beets, garden	20	19,733
Cabbages	9	8,024
Carrots	72	161,758
Cauliflower	80	25,326
Celery		
Cherries	16	6,696
Cucumbers	25	7,447
Mung beans		
Onion, bulb		
Papayas		
Passion fruit		
Peaches	17	32,725
Pears	64	89,126
Peppers		
Pimentos		
Plums		
Spinach	75	20,844
Sweet potatoes	90	52,052

Source: *Estimates of Dietary Oncogenic Risk, Regulating Pesticides in Food.* 1987. National Academy Press, Washington, D.C.

TABLE 11.5. Animal Feeds Not Subject to Processed Feed Regulations.

Alfalfa hay	Lespedeza hay	Rice straw
Almond hulls	Mullet forage (dry)	Safflower fodder
Clover	Mint hay	Sainfoin hay
Corn fodder	Oat hay, fodder, straw	Sorghum hay (fodder)
Cotton forage, by-products	Peanut hay	Soybean hay, straw
Cowpea fodder (dry vines)	Pea vine hay	Spearmint hay
Flax straw	Peppermint hay	Sugarcane fodder
Grass straw	Pigeon pea hay	Sunflower forage (dry)
Hops (dried)	Pineapple fodder (if dried)	Trefoil hay
Hop vines (dehydrated)	Rape straw	Vetch hay
Lentil hay	Rendered meat (cattle, poultry, swine, etc.)	Wheat straw

Source: *Estimates of Dietary Oncogenic Risk, Regulating Pesticides in Food.* 1987. National Academy Press, Washington, D.C.

rate of at least 50,000 kg annually. These herbicides are presented in Table 11.6, including year first tolerance granted, volume of use, and major crop uses.

Using EPA data and methods for describing oncogenic risk, more than 65% of oncogenic risk from all pesticides (insecticides, herbicides, fungicides) in meat, milk, dairy, and poultry products is due to herbicides. From the group of herbicides, more than 98% of all estimated oncogenic risk is due to the one herbicide, linuron. In beef, 99% of herbicide oncogenic risk is due to linuron with oxadiazon and alachlor contributing 0.4% each. The remaining herbicides contribute negligible risk. In pork, linuron is responsible for greater than 99% of the herbicide oncogenic risk. All other herbicides found in pork contribute negligible risk when compared to linuron.

The 10 herbicides with the highest estimated dietary oncogenic risk are

TABLE 11.6. Potentially Oncogenic Pesticides Identified by the EPA.

Active Ingredient (common/tradename)	Year Tolerance First Granted	Volume of Use (kg. active ingredient/yr.)	Major Crop Uses
Acifluorfen (Blazer®)	1980	633,484	Soybeans
Alachlor (Lasso®)	1969	38,461,538	Corn, soybeans
Arsenic acid	NA	NA	Cotton
Asulam	1975	NA	Sugarcane
Diallate	1969	226,244	Sugarbeets
Diclofop nethyl	1980	542,986	Soybeans
Ethalfluralin (Sonalan®)	1982	NA	Soybeans
Glyphosate (Roundup®)	1976	361,990	Hays, orchard crops
Linuron (Lorox®)	1966	3,167,420	Soybeans
Methanearsonic acid	NA	1,809,954	Cotton
Metolachlor (Dual®)	1976	17,239,819	Corn, soybeans
Oryzalin (Surflan®)	1974	769,230	Soybeans, vineyards
Oxadiazon (Ronstar®)	1977	NA	Rice
Paraquat (Gramoxone®)	1961	1,266,968	Rice, soybeans
Pronamide (Kerb®)	1972	45,248	Lettuce
Sodium arsenite	NA	NA	Grapes
Terbutryn wheat	1959	271,493	Barley
Trifluralin (Treflan®)	1963	17,647,058	Soybeans

NA = not available.
Source: *Estimates of Dietary Oncogenic Risk, Regulating Pesticides in Food.* 1987. National Academy Press, Washington, D.C.

TABLE 11.7. Estimated Oncogenic Risk from Dietary Exposure to Selected Herbicides.

Active Ingredient	Number of Food Uses	Relative Dietary Oncogenic Risk
Linuron	20	5568
Alachlor	25	88.5
Metolachlor	40	53
Oxadiazon	26	44
Oryzalin	57	42
Pronamide	25	28.5
Ethalfluralin	18	13
Diclofop methyl	5	7.5
Terbutryn	4	1.05
Glyphosate	134	1

Source: *Estimates of Dietary Oncogenic Risk, Regulating Pesticides in Food.* 1987. National Academy Press, Washington, D.C.

presented in Table 11.7. As illustrated by the table, the oncogenic risk of all approved herbicides is primarily due to a single chemical; linuron represents greater than 98% of all estimated risk of herbicides.

11.4.2 WHAT IS TO BE DONE?

Several conclusions may be made concerning the public health significance of herbicides in the food supply. First, the incidence of direct toxicity in humans following consumption of herbicide-sprayed crops or animals fed herbicide-sprayed crops is very low. Rapid decomposition and the use of less toxic chemicals has reduced the risk of toxicity. Meat, milk, and eggs from animals fed feedstuffs sprayed with chemical herbicides are not likely to present significant toxic risk to humans due to the short half-lives and rapid elimination of these herbicides from the animals' bodies. This is especially true of the newer, more water-soluble herbicides approved during the last 10 years.

Reducing the risk from oncogenic herbicides can probably be accomplished by reducing or eliminating exposure to a few high risk compounds. It has been estimated that elimination of limuron use alone would reduce the overall oncogenic risk from all pesticides in all foods by about 20%.

Introduction of new, nononcogenic, less toxic herbicides is proving to be an effective method to reduce public health risk. Several new herbicides were approved for use in 1986. The chemical structures of these compounds represented new classes in the industry, including the imidazolinones and the sulfonylureas. In addition to being nononcogenic, these herbicides are applied at lower rates than currently approved agents. Continued efforts to synthesize newer, more effective, and safer chemicals to support the \$2.7 billion U.S. herbicide market will promote high-yield production of crops while minimizing the risk to wildlife, domestic animals, and humans.

11.5 REFERENCES

Beste, C. E. 1983. *Herbicide Handbook,* Champaign Il: Weed Science Society of America, pp. 133, 443.

Clark, D. G., T. F. McElligott and E. W. Hurst. 1966. "The Toxicity of Paraguat," *Brit. J. Indust. Med.,* 23:126–132.

Environmental Protection Agency. 1982. *Pesticide Assessment Guidelines, Subdivision of Hazard Evaluation, Human and Domestic Animals,* Environmental Protection Agency, Office of Pesticides Programs.

Environmental Protection Agency. 1984. *Data Requirements for Pesticide Registration,* 49:42856–905.

Delvo, H. W. 1984. *Inputs: Outlook and Situation Report,* Washington, D.C.: Department of Agriculture, Economic Research Service.

Dover, M. J. 1985. *A Better Mousetrap: Improving Pest Management for Agriculture,* Washington, D.C.: World Resources Institute, p. 84.

Dudley, D. R. and J. R. Karr. 1980. "Water-Pesticides and PCB Residues in the Black Creek Watershed, Allen County, Indiana," *Pesticides Monitoring Journal,* 13:155–157.

Leng, M. L. 1972. "Residues in Milk and Meat and Safety to Livestock from the Use of Phenoxy Herbicides in Pasture and Rangeland," *Down to Earth,* 28:12–20.

Leng, M. L. 1977. "Comparative Metabolism of Phenoxy Herbicides," *Fate of Pesticides in Large Animals,* New York: Academic Press, pp. 54–76.

Mrak, E. M. 1974. "Chemistry and Analysis," *Herbicide Report,* Washington, D.C.: U.S. Environmental Protection Agency.

Oehler, D. D. and G. W. Ivie. 1980. "Metabolic Fate of the Herbicide Dicamba in a Lactating Cow," *J. Agric. Food Chem.,* 28:685–689.

Rizek, R. 1985. *Nationwide Food Consumption Survey, 1977–78.* Hyattsville, MD: U.S. Department of Agriculture.

Smalley, H. E. 1973. "Toxicity and Hazard of the Herbicide, Paraquat, in Turkeys," *Poultry Science,* 52:1625–1628.

Tshirley, F. H. 1987. "Weed Control," *Regulating Pesticides in Food,* Washington, D.C.: National Academy Press, pp. 228–234.

USFDA. 1986. *Annual Report of Food Residues in Domestic and Imported Commodities,* Washington, D.C.: U.S. Food and Drug Administration.

CHAPTER 12

Decontamination of Livestock

WILLIAM T. BUCK, D.V.M., M.S.[1]

12.1 INTRODUCTION

A true economic emergency exists for the livestock producer when violative residues occur in food animals. In addition to potential human health hazards, economic losses from a single contamination incident may extend to millions of dollars because of livestock and poultry death losses, additional laboratory expenses incurred by regulatory agencies and involved producers, loss of revenue because the producer cannot market the product, and legal fees associated with assessing liability.

In the past, most intoxications and contamination of livestock and poultry resulted from natural toxins (mycotoxins, phytotoxins, and minerals) in their feed supply. During the last four decades the development and widespread use of numerous synthetic compounds have provided new sources of chemical contaminants. For example, about 850 pesticides are registered with the U.S. Environmental Protection Agency (EPA). Their metabolites and other alteration products are not included in the register. In addition, the Office of Toxic Substances in the EPA has approximately 60,000 commercial chemicals in its registry (Jelinek, 1985).

Some of the common classes of chemicals and toxins that cause residue problems in food animals include:

(1) Pesticides (especially organochlorine insecticides and fungicides)

(2) Industrial effluents (especially metals and polyhalogenated hydrocarbons) (Osweiler et al., 1985)

---

[1]Professor of Veterinary Toxicology, Director of National Animal Poison Control Center, College of Veterinary Medicine, University of Illinois at Urbana, Illinois, USA.

(3) Feed additives (including growth promotants, antimicrobials and anthelmintics) (Hall, 1986)

(4) Natural toxins (especially certain mycotoxins) (Pier et al., 1980)

This chapter focuses upon the sources of chemicals and toxins that commonly cause violative residues in livestock, and recommended procedures for decontamination of exposed animals. Suggestions are presented for the prevention of such residues through feeding and management techniques. Case histories of two actual incidents of livestock contamination are presented, together with the methods employed for decontamination of the herds.

12.2 GENERAL CONSIDERATIONS

12.2.1 ACUTE VS. CHRONIC EXPOSURE

Acute exposures usually result from single, massive contamination of feed, often causing severe clinical signs and death of livestock. Chronic exposures may be repeated over a period of days, months or even years. They rarely cause sudden toxicosis, but rather insidious loss of body condition and reduced productivity. In either situation, residues in meat, milk and eggs may be a major concern.

Chronic exposures may be more difficult to diagnose than acute ones. Appropriate samplings of blood, urine and feces, and biopsies of fat and other tissues may be indicated. In addition, feed, water and environmental samples should be obtained and all specimens examined by chemical analysis or other tests for determining if exposure occurred. Management should be directed toward removing the source(s) of exposure, and preventing contamination of livestock feeds and pollution of the environment, as well as decontamination of exposed animals, feeds and the environment. Veterinarians, nutritionists and animal scientists may be involved in supervising the decontamination of food animals, disposal of unsalvageable animals, or slaughter of exposed animals.

12.2.2 MANAGEMENT OF EXPOSED ANIMALS

Animals manifesting signs of toxicosis should be treated, and efforts to prevent toxicosis in those that have been exposed should be initiated as soon as possible. Experience at the Illinois Animal Poison Information Center (Buck and Bralich, 1986) indicates that superactivated (i.e., greater than 3,000 M^2/g, dry weight) charcoal* is effective in preventing the ab-

*SuperChar-Vet, Gulf-Bio-Systems, Inc., 5310 Harvest Hill Rd., Dallas, TX 75203.

sorption of most organic compounds from the gastrointestinal tract. Because many xenobiotics and chemicals are excreted via the gastrointestinal (GI) tract, the oral administration of superactivated charcoal is indicated for animals that have been exposed dermally and systemically, as well as by the oral route, to prevent their reabsorption. Depending upon the rate of metabolism and excretion of the chemical(s) involved, repeated dosing with superactivated charcoal may be indicated.

Animals that have been contaminated by lipid soluble compounds, such as chlorinated pesticides and some industrial chemicals, may have violative residues in their body fat. Decontamination of the animal can be accelerated by promoting the loss of body fat through limited feeding, or increased butterfat production in milk, depending upon the management circumstances (Craigmill, 1986a). These techniques will be discussed later in this chapter.

12.3 POTENTIAL SOURCES

To list the potential chemicals and toxins that could contaminate livestock, and thus foods of animal origin, is beyond the scope of this chapter. Instead, a brief discussion of the common sources of food-animal contamination, together with certain of the most likely chemicals involved, is presented.

12.3.1 AGRICULTURAL CHEMICALS

The use of agricultural chemicals, especially pesticides and fertilizers, has resulted in contamination of livestock feeds and water supplies. The greatest hazards are associated with the use of insecticides and fungicides. The inappropriate storing and/or disposal of unused pesticides has frequently been a source of feed contamination. Too often, granular formations of pesticides appear similar to feed ingredients, and are inadvertently incorporated into animal feeds (Buck, 1973). Contamination of water supplies may occur from runoff of agricultural chemicals used on crops and soils.

12.3.2 INSECTICIDES

Contamination of livestock is usually accidental, resulting from excessive or nonrecommended exposure. Common sources include:

(1) Inadvertent mixing of granular or powdered insecticides into feeds, mistaken for salt or mineral preparations
(2) Storing loose grain or feed where an insecticide has previously been stored

(3) Formulated feeds prepared from contaminated ingredients, such as animal oils, meat scraps and silage by-products

(4) Use of persistent insecticides on soils where crops are grown for feeding to livestock

(5) The application of such persistent chemicals around animals and their housing, feed bins and elevators

(6) Applying nonrecommended formulations of compounds on animals

(7) Miscalculation or mismeasurement of insecticide formulations used on animals

The organochlorine insecticides are diffuse stimulants or depressants of the central nervous system. Since many (DDT, aldrin, dieldrin, heptachlor, chlordane, endrin, and mirex) are relatively persistent in the environment and in the adipose tissue, they are the most likely compounds to cause violative residues in foods of animal origin. Others, such as methoxychlor, lindane, and toxaphene are less persistent, and cause fewer residue problems (Osweiler et al., 1985).

The estimated half-life (time required for residue concentrations to reduce by half) for the more persistent compounds, such as DDT/DDE, dieldrin and heptachlor epoxide, varies with species, sex, stage of lactation and maturity of animals. Also, if a herd is exposed suddenly to very high doses (so that some animals die and others have residues of 50–100 ppm in their body fat), the rate of excretion is relatively rapid for the first 30 days (Osweiler et al., 1985). The author has experienced numerous incidents where massive exposures to aldrin has caused some deaths, and dieldrin residues in surviving animals have reached 100 ppm in milk and body fat. Within 30 days the residues have been reduced to 5–10 ppm. After the residue concentration is reduced to 5–10 ppm, the half-life increases drastically. A listing of adipose tissue or butterfat approximate half-lives (after the initial 30-day period) for these persistent insecticides in various animals is presented in Table 12.1.

When animals are contaminated with organochlorine insecticide residues, several procedures have been successful in accelerating removal of the residues from the body. These include starvation to remove body fat, and feeding activated charcoal intermittently over a several-week period. Each incident should be handled uniquely, however, depending upon the amount of residue, duration of exposure, and many management and economic factors. Because these compounds are readily excreted via milkfat, increasing the butterfat production of dairy cows will shorten the time necessary for decontamination.

The amount of weight reduction that will be effective in accelerating the elimination of residues will depend upon the condition of the animals. One

TABLE 12.1. Approximate Half-Lives of Persistent
Organochlorine Insecticides in Body Fat or Butterfat
of Various Types of Food Animals.

Type of Animal	Estimated Half-life (Days)
Finishing Steers (500–600 kg)	245
Finishing Heifers (500–600 kg)	85
Feeder Steers (150–300 kg)	180
Dairy Cow (full lactation)	60
Finishing Swine (100 kg)	60
Growing Swine (15–50 kg)	30
Poultry (adult hens, eggs)	35
Broilers	15

Source: Osweiler et al., 1985.

should attempt to reduce the carcass fat by 90% (Craigmill, 1986a). If the percentage of adipose tissue in the carcass can be estimated (usually from less than 10% up to 25% in the bovine species), one can estimate the amount of weight loss needed by subtracting 15% (bovine gastrointestinal contents) from the live body weight, then multiplying by the percent of estimated carcass fat. For example, for a 500 kg steer in good condition, subtract $500 \times .15 = 75$ kg, for a weight of 425 kg. If the animal has 20% carcass fat, then $425 \times .20 = 85$ kg. Then $.90 \times 85 = 76.5$ kg is the weight loss equivalent to 90% fat reduction.

This approach requires sampling and testing of representative animals in the contaminated herd to assess the progress of decontamination. Testing can be done by biopsy of adipose tissue from the perineal or scrotal areas. When the residue has been reduced to about 1 ppm (provided there has been 90% reduction in carcass fat), the animals can be refattened for slaughter. The dilution effect and concomitant excretion will reduce the residue to below the usual 0.3 ppm actionable level by the time they are finished.

The rate of decontamination of lactating dairy cows may be increased by promoting butterfat production, while they lose 45–70 kg of body weight (Craigmill, 1986b). Postparturient cows should be fed as usual for the first 7 days, i.e., high-energy feeds to decrease the chances of ketosis. From the 7th to 60th day, feed a low-energy, high-protein ration. Animals should be given 5–6 kg TDN 1–1.5 kg protein daily. The objective is to get them to lose 45–70 kg of body weight, and to mobilize body fat, thus increasing the amount of residue excreted in milk. Monitor cows and their milk production closely, because this regimen may make them susceptible to ketosis (Craigmill, 1986b). This treatment is especially recommended for cows fed contaminated whole dry feed. They will excrete residues in milk at higher levels after freshening because of increased mobilization of body fat.

Cows in mid to late lactation should be fed to optimize butterfat production—for example, 18% crude fiber and 88–90% TDN on a dry matter basis (Craigmill, 1986b). Feed each cow at least 2.5 kg of hay per day.

Calves born to cows fed contaminated feed are contaminated at the same level as the dam. These calves should not be used for veal.

12.3.3 INDUSTRIAL WASTES AND EFFLUENTS

Pollutants resulting from industrial operations are usually localized around a point source except in areas where multiple industries overlap. Hazards from industrial accidents are usually acute, and may involve widespread and lethal exposure of animals, plants and humans, as well as contamination of the environment. Usually, the immediate source of exposure will be either via air or water, followed by contamination of animal feed ingredients. Such incidents are relatively rare in comparison to chronic environmental pollution from industrial wastes and effluents.

The more common sources of industrial pollution are subtle and insidious, resulting from smokestack emissions that contaminate forages, and therefore the feed supply, or from contamination of water due to improper waste disposal. Pollutants that are most likely to be problems in livestock include: fluoride; arsenic; lead; molybdenum; copper; mercury; cadmium; halogenated hydrocarbons, such as polychlorinated (PCB) and polybrominated (PPB) biphenyls; dibenzodioxins (TCDD); petroleum products, such as crude oils, fuel oils, kerosene, and gasoline; triaryl phosphates; solvents, such as acetone, carbon disulfide, ethyl alcohol, isopropyl alcohol, methyl alcohol, turpentine, toluene, xylenes and petroleum naphthas (Osweiler et al., 1985).

Another frequent source of industrial pollution is the accident that results in misformulation of animal feeds or human food. Such well-known problems as contamination of animal feeds with fire retardant polybrominated biphenyls (PBB) in Michigan; polychlorinated biphenyl (PCB) contamination of rice oil and other lipid feed/food ingredients; and the use of trichloroethylene for extraction of soybean oil, resulting in a toxic soybean meal being incorporated into livestock feeds (Buck, 1979).

Polychlorinated biphenyls (PCB) and dioxins are the two most important environmental chemicals. PCB's are chemically and physically stable aromatic hydrocarbons that have been used as heat transfer agents in heating coils, electrical transformers and in sealer paint. The latter were used inside silos, resulting in contaminated livestock feeds. PCB's were also used as plasticizers, flame retardants, pressure sensitive adhesives, agents for reducing oxidation, to increase strength of fiberglass resins, and as chemical insulators (Osweiler et al., 1985).

In 1970, over 34 million kg were used in the U.S. alone. Their use has been banned, however, in the U.S. and several other countries. Their common name, aroclor, is a mixture of several compounds containing different concentrations of chlorine.

Compared to insecticides, PCB's are relatively low in toxicity, with several hundred ppm in the diet being required to cause overt toxicosis in most livestock species. The toxicity of PCB's increases with their chlorine content. Mink are very susceptible to PCB's, with birds, humans and livestock having a decreasing order of susceptibility (Osweiler et al., 1985). Feeding salmon and other animal scraps to mink, at 30 ppm PCB in their diet, has been associated with reproductive problems (Aulerich et al., 1973). Cattle have died after being exposed to back-rubber oil containing transformer oil (69–95% PCB) for 6 months (Robens and Anthony, 1980).

The ubiquitous use of PCB's has resulted in contamination of livestock feeds and human foods, as well as the environmental food chain. They are lipid soluble, being found in fats and oils, and are persistent in the body and the environment. Residues may be the greatest problem with PCB's in livestock. Body fat accumulation is roughly equivalent to (or up to several-fold higher than) the concentration of PCB's in the diet. The greater the chlorine content of the PCB, the greater the accumulation of residues. Feedlot cattle exposed to PCB back-rubber oil had up to 2,200 ppm PCB in their body fat (Robens and Anthony, 1980). PCB's are excreted in butterfat and eggs in a way similar to that of the most persistent organochlorine insecticides (Osweiler et al., 1985).

The rate of excretion of PCB's depends upon many factors, such as species, sex, age, amount of body fat and diet. The half-life of PCB's in livestock is comparable to that of dieldrin or heptachlor epoxide; i.e., up to 245 days in steers, 60 days in lactating cows and pigs, and 30–40 days in chickens. As with the organochlorine insecticides, decontamination of animals may be enhanced by reducing feed intake to accelerate removal of body fat. The intermittent feeding of superactivated charcoal may also accelerate decontamination.

Dioxins (chlorinated dibenzodioxins) are contaminants of several chemical products to which livestock and poultry have been exposed. These include the phenoxy herbicide, 2,4,5-T, hexachlorophenol (an antibacterial and antifungal agent), and pentachlorophenol (fungicidal wood preservative). The contamination process occurs during their manufacture (Osweiler et al., 1985). During the past several years, however, chemical manufacturers have cleaned up the manufacturing processes, resulting in reduced incidents of contamination.

Dioxins containing 4 chlorines (tetrachlorodibenzodioxin) are the most toxic to birds and mammals, with lethal doses being in the μg/kg range. Dioxins are cumulative, and continuous exposure to 1 μg/kg/day for 90 days will kill rats. The toxic effects may be delayed for weeks, months and perhaps years.

The half-life for excretion of dioxins has been reported to be 31 days in rats and 23–43 days in guinea pigs (Poiger and Schlatter, 1980; Ramsey, 1979). They are excreted via bile (feces) and urine, primarily as glucuronide metabolites (no free dioxin). Superactivated charcoal could be beneficial by absorbing dioxins or their conjugates in the GI tract, thus interrupting their enterohepatic recirculation.

Storage of industrial and municipal wastes in dumpsites that have been improperly constructed or maintained may result in the release of toxic chemicals into surrounding water supplies and the atmosphere. Such problems have caused public concern, and on certain occasions have been the subject of widespread news coverage. These problems are usually of a local nature, with animals and humans located near the point source being at greatest risk.

12.3.4 MYCOTOXINS

A mycotoxin, by definition, is a toxic metabolite of a fungus. Because of their widespread natural occurrence in animal feedstuffs and human foods, the importance of mycotoxins has been recognized in such fields as commerce, law and politics, as well as in animal and human health. Because of the potential carcinogenic and other public health effects of residues in meat, milk and eggs, producers should be aware of those mycotoxins that are most likely to present problems. Metabolism studies have been performed in animals using aflatoxin B, ochraloxin A, patulin, penicillic acid, sterigmatocystin, T-2 toxin, diacetoxyscirpenol (DAS), deoxynivalenol (DON), zearelenone and several other mycotoxins. Of these, aflatoxin is the most likely to cause residues in meat, milk and eggs, mostly in the forms of aflatoxin M or B. Ochratoxin A residues have been reported in swine tissues. None of the other mycotoxins have been shown to cause serious residue problems in foods of animal origin.

Many animals, especially ruminants, are excellent detoxifiers of mycotoxins. For instance, the diet:milk ratio for aflatoxin in dairy cows is at nearly 300:1 (Pier et al., 1980). The diet:skeletal muscle ration of aflatoxin ($B_1 + M_1$) of cattle varies from 250 to 14,000:1. Of the edible tissues, liver and kidneys are the most likely to contain aflatoxin residues. In situations where a mycotoxin residue has been found in milk or animal tissues, the regulatory attitude has been one of prudence. In the U.S., an action level of 0.5 ppb has been set for aflatoxin M_1 in milk. In Denmark, all suspect swine kidneys are analyzed and, if ochratoxin is found, the entire carcass is condemned (Pier et al., 1980).

Animal health specialists may be asked to assist producers whose livestock have been exposed to contaminated grain. When aflatoxin is involved, all dietary sources of aflatoxin B_1, and milk or tissue concentrations of aflatoxin M_1, should be identified and quantitated. The calculated diet:milk or diet:tissue ratio can be used to determine the maximum amount of aflatoxin-contaminated feed that can be given without resulting in actionable residues in the food products. Dairy cows whose milk has above actionable levels of aflatoxin should be placed on a clean ration for 3–5 days to facilitate decontamination. During this period, their milk should not be used for human consumption. In general, however, if mycotoxin-contaminated grain must be fed to food animals, it should be given to mature, nonlactating or nonpregnant animals that will not be slaughtered for 30 days. Even so, because of their detoxification capabilities, it is more appropriate to feed mycotoxin contaminated grain to livestock than to allow it to enter the human food chain.

12.3.5 FEED ADDITIVES AND ANTIBACTERIAL AGENTS

Feed additives are an integral part of livestock production. Some type of additive is included routinely in many gestation rations, nursery feeds, and growing-finishing rations. Growth promotants added to livestock rations include such antibiotics as the tetracyclines, lincomycin, neomycin, bacitracin, the bambermycins, erythromycin, oleandomycin, penicillin, streptomycin, tylosin and virginiamycin. Nonantibiotic growth promotants include sulfamethazine, sulfathiazole, sodium arsanilate, furazolidone, nitrofurazone, roxarsone, and carbadox. Other important feed additives include various dewormers, such as thiabendazole, hygromycin B, levamisole, pyrantel tartrate, dichlorvos, coumaphos, phenothiazine, piperazine and fenbendazole. The U.S. Food and Drug Administration (FDA) has established tissue residue tolerances and preslaughter withdrawal times for most feed additives. Hall (1986) has published a summary of residue tolerances and withdrawal times for feed additives used in pork production.

Sulfamethazine is the most commonly identified violative residue in pork tissue, probably because it is the most widely used feed additive in pork production (Hall, 1986). Considerable research has been done to identify the sources of such violative residues. The sources and their rates of incidence have included (Hall, 1986):

(1) 57% – contamination of withdrawal feed with sulfonamide (e.g., feeders not cleaned when switching from medicated to nonmedicated feed)

(2) 25% — withdrawal time not observed
(3) 12% — accidents, e.g., finishing hogs breaking into medicated feed
(4) 2.4% — recycling of feces and urine
(5) 2% — unapproved use of drug
(6) 2% — sulfonamide contamination of watering system

Therefore, it is probable that if feed formulation recommendations and established withdrawal times are strictly followed, there will be no violative residues of feed additives.

12.4 FIELD CASES

12.4.1 ALDRIN/DIELDRIN CONTAMINATION OF DAIRY CATTLE

In 1983, a Missouri dairyman cleaned a storage bin of various debris, waste grain and broken bags of chemicals, which consisted of an unknown amount of aldrin insecticide. The mixture of waste material was dumped in a rubbish heap located in a pasture where 23 holstein dry cows and pregnant heifers were located. Shortly thereafter the cattle invaded the rubbish heap and consumed the contaminated debris and grain. During the following week, over half the animals exhibited signs of organochlorine insecticide toxicosis: intermittent convulsive seizures, muscular twitching, frenzied activity and abnormal posturing, followed by central nervous system depression. Three animals died and the remainder of them appeared fully recovered after 7 days. All of the exposed animals freshened during the ensuing 2 months; 5 within 2 weeks of the exposure. Colostrum and butterfat samples from these 5 animals contained from 50 to 100 ppm dieldrin and 0.5 to 3 ppm aldrin (aldrin is epoxidized to dieldrin in the animal body). Colostrum and butterfat samples taken from those cows that freshened 6–8 weeks after exposure contained only 5–10 ppm dieldrin. The dairyman elected to attempt decontamination of the 20 exposed cows, rather than destroy them, as is often the case. His lactating herd consisted of about 100 cows, who were routinely turned on pasture during the daytime. The 20 contaminated cows were kept separate from the noncontaminated herd, and they were milked last. Their milk was saved and fed to young calves. As cows freshened and additional milk was produced, additional heifer calves were purchased from various sources, in order to utilize the milk that contained violative residues of dieldrin. Individual milk samples from the exposed cows were analyzed about every 2 months, and those cows having less than 0.2 ppm dieldrin in their butterfat were returned to the milking herd. Samples of bulk milk from the clean herd were also analyzed every 2 months. The calves fed the contaminated milk through 3–4 months of age were maintained separate from other animals, and were eventually bred to parturate when they were 2 to 2½ years of age. Fat biopsy samples for dieldrin analysis were surgically taken from the perineal area of the pregnant heifers when they were 18–24 months of age.

The results of this decontamination management were very acceptable to the dairyman. All 20 cows that had ingested the aldrin were returned to the clean herd during the first lactation period subsequent to exposure. At no time was the dieldrin concentration greater than 0.1 ppm in the bulk milk butterfat from the clean herd. Over 80 very nice heifers were raised on the contaminated milk, and none of them had greater than 0.1 ppm dieldrin in their body fat at 18 to 24 months of age. These heifers were used as replacements, or were sold to other dairymen. The dairyman commented that the net cost of his contamination problem consisted only of the death of the three cows associated with the initial aldrin exposure.

12.4.2 AFLATOXIN CONTAMINATION OF SEVEN DAIRY HERDS

In January 1981, aflatoxin M_1 was detected in silo milk from 2 of 14 routes in southern Illinois at concentrations in excess of 1 ppb. Silo milk from a third route contained greater than 0.5 ppb. These three routes consisted of producers. Bulk milk samples from each producer were analyzed and seven of the 29 were found to contain actionable levels of aflatoxin M_1 (in excess of .5 ppb). Milk from each of the seven producers was ordered to be discarded until it became free of aflatoxin. These problem herds were located in an area where drought conditions in 1980 had seriously reduced corn production. Scientists at the University of Illinois were asked to consult with the producers, veterinarians, and representatives of the milk cooperative regarding feeding management and decontamination procedures with the problem herds.

The aflatoxin contamination in the seven herds was confirmed on a Thursday. Each producer was notified and requested to submit samples of all feed being given to the lactating cows to the Illinois Department of Agriculture Laboratory, Centralia, IL, for aflatoxin analysis. By Friday evening, results of the analyses were available and the consulting team was visiting the farms to assist the producers. Because a significant amount of milk was being discarded, the producers were anxious to implement management procedures necessary for decontamination. Armed with the aflatoxin data on corn and feed samples submitted by the seven producers, the team proceeded to counsel each producer in determining the most judicious procedure for getting his milk free of aflatoxin and back on the market.

It was documented that all except one of the producers were feeding corn grain that was grown on their farm in 1980. In most cases, they were also feeding corn silage that was produced in 1980. The silage usually was made with the corn that had been damaged the greatest by the drought. Chemical analysis of several samples of corn silage revealed no detectable aflatoxin. The concentrations of aflatoxin B_1 in the various corn samples are presented in Table 12.2.

In each case the producer had access to noncontaminated corn raised in 1979, or was able to purchase replacement corn or complete feed concentrate to get the cows on a diet free of aflatoxin for a period of 72 hours. The objective was to subsequently reduce the aflatoxin content of the concentrate to less than 100 ppb by blending their contaminated corn with noncontaminated corn. This recommendation was based upon the premise that the dairy cow can degrade dietary aflatoxin; that the concentrate:milk ratio (i.e., aflatoxin B_1 in concentrate to M_1 in milk) is about 300:1.

On the next Monday (after 72 hours on noncontaminated diets) bulk milk samples from the seven producers were analyzed and found free of aflatoxin, or well below the action level. An interesting side note was that one producer reported that milk production increased by 2.5 kg/cow after switching to a concentrate formulated with noncontaminated corn.

Milk from these and neighboring herds was monitored for aflatoxin content during the subsequent several months. Aflatoxin M_1 content remained below the actionable level in all herds where the concentrate fed contained no more than 100 ppb aflatoxin B_1.

12.5 SUMMARY

The most common sources of violative residues in food animals include organochlorine insecticides and fungicides, polychlorinated biphenyls,

TABLE 12.2. Performance Data of Each Dairy Herd and Amount of Aflatoxin Contamination of Corn.

Herd Number	No. Cows Milked	Daily Milk Prod. (kg)	Contaminated Corn (bu)	Aflatoxin in Corn (ppb)
1	75	1,455	9,000	850
2	100	1,850	2,000	800
3	40	1,000	2,000	200
4	42	1,090	2,000	200
5	49	Not obtained	Not obtained	Not obtained
6	56	Not obtained	4,000	240
7	91	Not obtained	4,000	700

Source: Buck et al., 1981.

dibenzodioxins, certain antibacterial feed additives and therapeutic agents, and mycotoxins (especially aflatoxin). In addition to potential human health hazards, economic losses from a single contamination incident may extend to millions of dollars because of livestock and poultry death losses, additional laboratory expenses incurred by regulatory agencies and producers, loss of revenue because the producer cannot market the product, and legal fees associated with assessing liability. Suggestions are presented for the prevention of such residues through feeding and management techniques. The decontamination of animals with violative residues is also discussed. Techniques used for accelerating the natural processes of decontamination by the body include forced reduction of body fat, dilution of contaminated grain by blending, and oral administration of superactivated charcoal to interrupt enterohepatic recirculation.

12.6 REFERENCES

Aulerich, R. J., R. K. Ringer and S. Iwamoto. 1973. "Reproductive Failure and Mortality in Mink Fed on Great Lakes Fish," *J. Reprod. Fert. Suppl.*, 19:377.

Buck, W. R. 1969. "Pesticides and Economic Poisons in the Food Chain," *Proc. 73rd Ann. Meet. U.S. An. Health Assn.*, pp. 221–226.

Buck, W. R. 1979. "Animals as Monitors of Environment Quality," *Vet. Human Toxicol.*, 21(4):277.

Buck, W. B. 1973. "Color Identification of Granular Insecticide Formulations," in *Pesticides and the Environment: A Continuing Controversy*, W. B. Diechmann, ed., New York: Intercontinental Medical Book Corp., pp. 565–568.

Buck, W. B. and P. M. Bralich. 1986. "Activated Charcoal: Preventing Unnecessary Death by Poisoning," *Vet. Med.* (Jan):73.

Buck, W. B. and W. VanNote. 1968. "Aldrin Poisoning Resulting in Dieldrin Residues in Meat and Milk," *J. Amer. Vet. Med. Assoc.*, 1553:1472.

Craigmill, A. L. 1986a. "Heptachlor Contamination of Livestock and Poultry," *Heptachlor Residue Education Task Force Fact Sheet*, USDA Extension Service.

Craigmill, A. L. 1986b. "Heptachlor Contamination of Dairy Cattle," *Heptachlor Residue Education Task Force Fact Sheet*, USDA Extension Service.

Hall, H. F. 1986. "Residue Avoidance in Pork Production," *Compend Cont. Ed.*, 8(4):200.

Jelinek, C. F. 1985. "Control of Chemical Contaminants in Foods: Past, Present and Future," *J. Assoc. Off. Anal. Chem.*, 68:1065.

Osweiler, G. D., T. L. Carson, W. B. Buck and G. A. VanGelder. 1985. *Clinical and Diagnostic Veterinary Toxicology*, 3rd ed., Dubuque, IO: Kendall/Hunt Pub.

Pier, A. C., J. L. Richard and S. J. Cysewski. 1980. "Implications of Mycotoxins in Animal Disease," *J. Amer. Vet. Med. Assoc.*, 176:719.

Richard, J. L., A. C. Pier, R. D. Stubblefield, O. L. Shotwell, R. L. Lyon and R. C. Cutlip. 1983. "Effect of Feeding Corn Naturally Contaminated with Aflatoxin on Feed Efficiency, on Physiologic, Immunologic and Pathologic Changes, and on Tissue Residues in Steers," *Amer. J. Vet. Res.*, 44:1294.

Poiger, H. and Schlatter C. 1980. "Influence of Solvent and Adsorbents on Dermal and Intestinal Absorption of TCDD," *Food Cosmet. Toxicol.*, 18:477.

Ramsey, J. C. 1979. "The In Vivo Biotransformation of 2,3,7,8 Tetrachlorodibenzo-*p*-dioxin in the Rat," *Toxicol. Appl. Pharmicol.*, 36:209.

Robens, J. and H. D. Anthony. 1980. "Polychlorinated Biphenyl Contamination of Feeder Cattle," *J. Amer. Vet. Med. Assoc.*, 177:613.